Annals of Mathematics Studies

Number 118

UNITARY REPRESENTATIONS OF REDUCTIVE LIE GROUPS

BY

DAVID A. VOGAN, JR.

PRINCETON UNIVERSITY PRESS

———

PRINCETON, NEW JERSEY

1987

The Annals of Mathematics Studies are edited by
William Browder, Robert P. Langlands, John Milnor, and Elias M. Stein
Corresponding editors:
Stefan Hildebrandt, H. Blaine Lawson, Louis Nirenberg, and David Vogan, Jr.

Clothbound editions of Princeton University Press
books are printed on acid-free paper, and binding
materials are chosen for strength and durability. Pa-
perbacks, while satisfactory for personal collections,
are not usually suitable for library rebinding

ISBN 0-691-08481-5 (cloth)
ISBN 0-691-08482-3 (paper)

Printed in the United States of America
by Princeton University Press, 41 William Street
Princeton, New Jersey

☆

Library of Congress Cataloging in Publication data will
be found on the last printed page of this book

To my parents, for keeping the first draft

CONTENTS

ACKNOWLEDGEMENTS

This book is based very loosely on the Hermann Weyl
Lectures given at the Institute for Advanced Study in
Princeton, in January, 1986. I am very grateful to Robert
Langlands and the Institute for the invitation to deliver
those lectures, and for their hospitality during my visit.
During 1985 and 1986, there was a seminar at MIT on unipotent representations; I am grateful to all the participants
for their unflagging dedication to mathematical truth, and
for staying awake most of the time. Susana Salamanca-Riba
and Jeff Adams have examined parts of the manuscript, and
provided helpful advice throughout its writing. Tony Knapp
made detailed comments on the entire manuscript; in those
few places where I did not follow his advice, the reader
will probably wish I had.

Most of all, I must thank Dan Barbasch. Many of the
ideas on unipotent representations discussed here are joint

work, and all of them are influenced by the deep results he has obtained on unitary representations of classical groups. Our collaboration is a continuing mathematical and personal pleasure.

This work was supported in part by grants from the National Science Foundation and the Sloan Foundation.

Unitary Representations of Reductive Lie Groups

INTRODUCTION

Perhaps the most fundamental goal of abstract harmonic analysis is to understand the actions of groups on spaces of functions. Sometimes this goal appears in a slightly disguised form, as when one studies systems of differential equations invariant under a group; or it may be made quite explicit, as in the representation-theoretic theory of automorphic forms. Interesting particular examples of problems of this kind abound. Generously interpreted, they may in fact be made to include a significant fraction of all of mathematics. A rather smaller number are related to the subject matter of this book. Here are some of them.

Let X be a pseudo-Riemannian manifold, and G a group of isometries of X. Then X carries a natural measure, and G acts on $L^2(X)$ by unitary operators. Often (for example, if the metric is positive definite and complete) the Laplace-Beltrami operator Λ on X is self-adjoint. In that case, G will preserve its spectral

3

decomposition. Conversely, if the action of G is transitive, then any G-invariant subspace of $L^2(X)$ will be preserved by Δ. The problem of finding G-invariant subspaces therefore refines the spectral problem for Δ.

The prototypical example of this nature is the sphere S^{n-1}, with G the orthogonal group $O(n)$. If n is at least 2, the minimal invariant subspaces for $O(n)$ acting on $L^2(S^{n-1})$ are precisely the eigenspaces of the spherical Laplacian. (This is the abstract part of the theory of spherical harmonics.) If n is 2, we are talking about Fourier series. The fundamental importance of these is clear; but they may of course be analyzed without explicit discussion of groups. For $n = 3$, the theory of spherical harmonics leads to the solution of the Schrödinger equation for the hydrogen atom. Here the clarifying role of the group is less easy to overlook, and it was in this connection that the "Gruppenpest" entered quantum physics in an explicit way.

A second example, still in the framework of pseudo-Riemannian manifolds, is the wave operator. Viewed on a four-dimensional space-time manifold, this is just the Laplace-Beltrami operator for a metric of signature $(3,1)$. If the manifold has a large isometry group (for instance, if it is Minkowski space), then the space of solutions can often be described terms of this group action.

An example with a rather different flavor is the space X of lattices (that is, discrete subgroups isomorphic to $\mathbb{Z}^n$) in $\mathbb{R}^n$. An automorphic form for $G = GL(n,\mathbb{R})$ is a smooth function on X, subject to some technical growth and finiteness conditions. (Actually it is convenient to consider at the same time various covering spaces of X, such as (for fixed p) the space of lattices L endowed with a basis of L/pL.) It is easy to imagine that functions on X have something to do with number theory, and this is the case. One goal of the representation-theoretic theory of automorphic forms is to understand the action of G on the space of automorphic forms. Because the G-invariant measure on X has finite total mass (although X is not compact), this problem is closely connected to the corresponding L^2 problem. An introduction to this problem may be found in [Arthur, 1979].

Finally, suppose X is a compact locally symmetric space. (Local symmetry means that $-Id$ on each tangent space exponentiates to a local isometry of X. An example is a compact Riemann surface.) We seek to understand the deRham cohomology groups of X. Here there is no group action in evidence, and no space of functions. However,

Hodge theory relates the cohomology to harmonic forms on X,

so the latter defect is not serious. For the former, we

consider the bundle Y over X whose fiber at p is the

(compact) group K_p of local isometries of X fixing p.

Harmonic forms on X pull back to Y as certain vector-

valued functions. On the other hand, Y has a large

transitive group G acting on it. (G may be taken to be

the isometry group of the universal cover of X; Y is then

the quotient of G by the fundamental group of X.) The

cohomology of X can now be studied in terms of the action

of G on functions on Y. Perhaps surprisingly, this has

turned out to be a useful approach (see [Borel-Wallach,

1980]).

With these examples in mind, we recall very briefly the

program for studying such problems which had emerged by 1950

or so. The first idea was to formalize the notion of group

actions on function spaces. In accordance with the general

philosophy of functional analysis, the point is to forget

where the function space came from.

Definition 0.1. Suppose G is a topological group. A *repre-*

sentation of G is a pair (π, V) consisting of a complex

topological vector space V, and a homomorphism π from G

to the group of automorphisms of V. We assume that the map
from G×V to V, given by

$$(g,v) \to \pi(g)v$$

is continuous. An *invariant subspace* of the representation
is a subspace W of V which is preserved by all the operators $\pi(g)$ (for g in G). The representation is called
reducible if there is a closed invariant subspace W other
than V itself and {0}. We say that π is *irreducible* if
V is not zero, and π is not reducible.

The problem of understanding group actions on spaces of
functions can now be formalized in two parts: we want first
to understand how general representations are built from
irreducible representations, and then to understand irreducible representations. This book is concerned almost exclusively with the second part. Nevertheless, we may hope to
gain a little insight into the first part along the way,
much as one may study architecture by studying bricks.

If we take G to be $\mathbb{Z}$, then a representation is determined by a single bounded invertible operator, $\pi(1)$. The
only interesting irreducible representations of G are the
one-dimensional ones (sending 1 to a non-zero complex number). The decomposition problem in this case amounts to
trying to diagonalize the operator $\pi(1)$. There are some

things to say about a single operator; but on infinite-dimensional spaces, one needs more hypotheses to begin to develop a reasonable theory. The easiest assumption to use is that the space is a Hilbert space and that the operator commutes with its adjoint. (Such operators are called normal.) In that case, the space can be decomposed in some sense into an "integral" of eigenspaces of the operator.

Once we decide to focus on group actions on Hilbert spaces, it is easy to see how these could arise naturally in our original problem. If G acts in a measure-preserving way on a measure space X, then it acts by unitary operators on $L^2(X)$. The continuity condition in the definition of representation comes down to this: if S is a subset of X of finite measure, and g is a small element of G, then g.S differs from S only in a set of small measure. This is clear for smooth Lie group actions on manifolds with smooth densities. The following definition therefore admits many examples.

Definition 0.2. The representation $(\pi, \mathcal{H})$ is called *unitary* if $\mathcal{H}$ is a Hilbert space and the operators $\pi(g)$ are unitary (that is, they preserve the inner product in $\mathcal{H}$).

All of the L^2 spaces mentioned in the examples provide examples of unitary representations.

If $\mathscr{S}$ is a closed invariant subspace of the unitary representation $(\pi, \mathscr{H})$, then the orthogonal complement $\mathscr{S}^\perp$ of $\mathscr{S}$ is also invariant, and

$$\mathscr{H} = \mathscr{S} \oplus \mathscr{S}^\perp.$$

We would like to iterate this process, and finally write $\mathscr{H}$ as a direct sum of irreducible unitary representations. To see why this is not possible, take G to be $\mathbb{R}$, acting by translation on $L^2(\mathbb{R})$. Any invariant subspace $\mathscr{S}$ of $\mathscr{H}$ corresponds to a measurable subset S of $\mathbb{R}$, by

$$\mathscr{S} = \{f \in L^2(\mathbb{R}) \mid \hat{f} \text{ vanishes almost everywhere outside } S\}.$$

(Here $\hat{f}$ denotes the Fourier transform of f.) Since the measure space $\mathbb{R}$ has no atoms, it follows that any non-zero invariant subspace of $L^2(\mathbb{R})$ has a proper invariant subspace.

On the other hand, the Fourier transform in this example does exhibit $L^2(\mathbb{R})$ as a sort of continuous (or measurable) direct sum of translation invariant "subspaces," consisting of functions with Fourier transform supported at a single point ξ. These spaces are one-dimensional (consist-

ing of multiples of the function exp(ixξ)) and therefore
irreducible.

A fundamental theorem, going back to [Mautner, 1951],
guarantees the existence of such a decomposition in great
generality. The proof is based on von Neumann's theory of
rings of operators.

THEOREM 0.3 (cf. [Dixmier, 1981]). *Let G be a type* I
separable locally compact group, and let $(\pi,\mathcal{H})$ *be a*
unitary representation of G. *Then* π *may be written*
uniquely as a direct integral of irreducible unitary repre-
sentations of G.

For the definitions of direct integral and type I, we refer
to [Dixmier, 1981]. All Lie groups (even over local fields)
with countably many connected components are separable and
locally compact. Type I Lie groups include nilpotent, reductive, and algebraic ones. Examples of groups not of type I
are free groups on more than one generator, and certain solvable Lie groups.

There is no analogous general theorem for decomposing
non-unitary representations; yet these may be of the most
direct interest in applications. (We are rarely content to
know only that a solution of a differential equation exists

in L^2). Fortunately, L^2 harmonic analysis often provides
a guide and a tool for studying other function spaces, like
L^p or C^∞. We will not pursue this topic further; one
place to begin to look is in [Helgason, 1984].

Because of the theorem of Mautner, it is of particular
interest to understand the irreducible unitary representations of a group G.

Definition 0.4. Suppose G is a topological group. The
set of equivalence classes of irreducible unitary representations of G is written $\hat{G}_u$.

We will later use the notation $\hat{G}$ for a certain larger
class of irreducible representations of a reductive Lie
group G.

The irreducible unitary representations of $\mathbb{R}$ are the
characters of $\mathbb{R}$, the continuous homomorphsms of $\mathbb{R}$ into
the circle. These are just the functions of the form
$$x \to e^{ixy}$$
with y real. With proper hindsight, the characters of finite abelian groups can be found in classical number theory.
By the late nineteenth century, Frobenius and Schur were
beginning to study the irreducible representations of non-
abelian finite groups. In the 1920's, Weyl extended their

ideas to compact Lie groups ([Weyl, 1925], [Peter-Weyl,

1927]). (In fact, because of the simple structure of com-

pact groups, Weyl's results in that case were substantially

more complete than those available for finite groups. Even

today, the irreducible representations of compact connected

Lie groups are far better understood than those of finite

simple groups.) Weyl's work is in many respects the begin-

ning of the representation theory of reductive groups; parts

of it are summarized in Chapter 1.

In the 1930's, the representation theory of noncompact

groups began to be studied seriously. The books [Pontriagin,

1939] and [Weil, 1940] each contain (among other things) a

rather complete treatment of the unitary representations of

locally compact abelian groups. At the same time, quantum

mechanics suggested the problem of studying the Heisenberg

group H_n. This is the $2n+1$-dimensional Lie group of $n+1$

by $n+1$ matrices having 1's on the diagonal, and 0's every-

where else except in the first row and the last column. H_n

is nearly abelian: its commutator subgroup coincides with

its one-dimensional center. Its unitary representations

were determined completely by the Stone-von Neumann theorem

(cf. [von Neumann, 1931]).

In [Wigner, 1939], the physicist Eugene Wigner (still

motivated by quantum mechanics) made a study of the irredu-

cible representations of the Poincaré group. This is the
group of orientation-preserving isometries of (the pseudo-
Riemannian manifold) $\mathbb{R}^4$ with the Lorentz metric. He found
all of those which were of interest to him, but did not succeed in obtaining a complete explicit description of $\hat{G}_u$.
The missing representations he constructed in terms of the
irreducible unitary representations of two subgroups of the
Poincaré group: the Lorentz group $SO(3,1)$ (the isotropy
group of the origin in $\mathbb{R}^4$), and its subgroup $SO(2,1)$.

Wigner's analysis was based on studying the restriction
of a unitary representation to normal subgroups. The reason
that it could not succeed for $SO(2,1)$ and $SO(3,1)$ is
that these groups (or rather their identity components) are
simple. Within a few years, their irreducible unitary representations had been determined, in [Bargmann, 1947] and
[Gelfand-Naimark, 1947]. We will return to this branch of
the development in a moment.

Around 1950, Mackey extended Wigner's methods enormous-
ly, into a powerful tool ("the Mackey machine") for studying
unitary representations of G in terms of those of a normal
subgroup N of G, and of the quotient G/N. Some of his
work may be found in [Mackey, 1976]. His results were applied to successively larger classes of groups over the next
thirty years; two high points are [Kirillov, 1962] and

[Auslander-Kostant, 1971]. For algebraic Lie groups, this study was brought to a fairly satisfactory culmination by the following theorem of Duflo. (Notice that it is a direct generalization of the result of Wigner for the Poincaré group.)

THEOREM 0.5 ([Duflo, 1982]). *Let G be an algebraic group over a local field of characteristic 0. Then the irreducible unitary representations of G may be explicitly parametrized in terms of those of certain reductive subgroups of* G.

This is a slight oversimplification, but the precise statement is as useful as (and more explicit than) this one. There is a similar statement for all type I Lie groups, involving semisimple Lie groups instead of reductive ones. Some problems remain - for example, the explicit construction of the representations is not as satisfactory as their parametrization - but to some extent the study of $\hat{G}_u$ is reduced to the case of reductive groups. Here the subject matter of the book itself really begins. We will conclude the introduction with a quick outline of the material to be discussed.

    We begin with a definition.

Definition 0.6. A Lie group G is called *simple* if

    a) dim G is greater than one;

    b) G has only finitely many connected components;

    c) any proper normal subgroup of the identity

component of G is finite.

We say that G is *reductive* if it has finitely many connected components, and some finite cover of the identity

component G_0 is a product of simple and abelian Lie

groups. It is *semisimple* if there are no abelian factors in

this decomposition. Finally, G is said to be of *Harish-
Chandra's class* if the automorphisms of the complexified Lie

algebra defined coming from Ad(G) are all inner.

This definition of reductive is chosen to please no one. It

does not include all the semisimple groups needed in Duflo's

theorem for non-algebraic groups; but it is significantly

weaker (in terms of the kind of disconnectedness allowed)

than most of the definitions generally used.

    The lowest dimensional noncompact simple group is

$SL(2,\mathbb{R})$, the group of two by two real matrices of determinant one. It is a two-fold cover of the identity component

of SO(2,1) and its irreducible unitary representations

were determined in the paper [Bargmann, 1947] mentioned

earlier. We will not give a detailed account of his work;
this may be found in many places, including [Knapp, 1986]
and [Taylor, 1986]. The answer, however, contains hints of
much of what is now known in general. We will therefore
give a qualitative outline of it, as a framework for describing the contents of the book.

First, there are two series of representations with an
unbounded continuous parameter (the *principal series*). The
representations are constructed by real analysis methods.
An appropriate generalization, due to Gelfand and Naimark,
is in Chapter 3. These representations tend to contribute
to both the discrete and the continuous parts of direct
integral decompositions of function space representations.

Second, there is a series of representations with a
discrete parameter (the *discrete series*). These were originally constructed by complex analysis. They can contribute
only as direct summands in direct integral decompositions.
Harish-Chandra found an analogue of this series for any
reductive group in [Harish-Chandra, 1966]; this paper is one
of the great achievements of mathematics in this century. A
brief discussion of its results appears in Chapter 5. The
most powerful and general construction of such representations now available is an algebraic analogue of complex

analysis on certain homogeneous spaces, due to Zuckerman.
It is presented in Chapter 6.

Third, there is a series with a bounded continuous parameter (the *complementary series*). They are constructed from
the principal series by an analytic continuation argument.
One hopes not to need them for most harmonic analysis problems; for automorphic forms on $SL(2,\mathbb{R})$, this hope is called
the Ramanujan-Petersson conjecture. The construction of complementary series for general groups will be discussed in
Chapter 4; but we will make no attempt to be as general as
possible. Complementary series constructions have been
intensively investigated for many years (see for example
[Baldoni-Silva-Knapp, 1984] and the references therein), but
the precise limits of their applicability are still not at
all clear.

Finally, there is the trivial representation. For a
general reductive group, the trivial representation belongs
to a finite family of "unipotent representations." In addition to a name, we have given to these objects a local habi-
tation in Chapters 7 through 12. All that they lack is a
complete definition, a reasonable construction, a nice general proof of unitarity, and a good character theory. (More
information can be found in [S, 1594].)

The latter part of the book is largely devoted to a
search for a definition of "unipotent." A vague discussion
of what is wanted appears in the Interlude preceding Chapter
7. Chapters 7 through 11 discuss various subjects closely
related to the representation theory of reductive groups, to
see what they suggest about the definition. Chapter 12 is a
short summary of some of the main results of the search; it
contains a partial definition of unipotent.

Implicit in this discussion is the hope that the ideas
described here suffice to produce all the irreducible unitary representations of any reductive group G. Because the
constructions of complementary series and unipotent representations are still undergoing improvement, this hope is as
yet not precisely defined, much less realized. Nevertheless, a wide variety of partial results (for special groups
or representations) is available. Some of these are discussed in Chapter 13. I hope that the reader will be not
disappointed by this incompleteness, but enticed by the work
still to be done.

Chapter 1

COMPACT GROUPS AND THE BOREL-WEIL THEOREM

Our goal in this chapter is to recall the Cartan-Weyl description of the irreducible unitary representations of compact Lie groups, together with the Borel-Weil realization of these representations. A good general reference for the material is [Wallach, 1973]. The absence of more specific references for omitted proofs in this chapter is not intended to indicate that the results are obvious.

We begin with a little general notation. If G is a Lie group, we will write

$$G_0 = \text{identity component of } G$$

$$\mathfrak{g}_0 = \text{Lie}(G) \quad \text{(the Lie algebra)}$$

(1.1) $$\mathfrak{g} = \mathfrak{g}_0 \otimes_{\mathbb{R}} \mathbb{C} \quad \text{(the complexification of } \mathfrak{g}_0)$$

$$U(\mathfrak{g}) = \text{universal enveloping algebra of } \mathfrak{g}$$

$$(\mathfrak{g}_0)^* = \text{real-valued linear functionals on } \mathfrak{g}_0$$

$$\mathfrak{g}^* = \text{complex-valued linear functionals on } \mathfrak{g}_0.$$

The circle group $\mathbb{T}$ of complex numbers of absolute value 1 has Lie algebra $i\mathbb{R}$. With this identification, the exponential map for $\mathbb{T}$ is just the usual exponential function. We often identify $\mathbb{T}$ with the unitary operators on a one-dimensional Hilbert space. Similarly, the multiplicative group $\mathbb{C}^{\times}$ of non-zero complex numbers has Lie algebra $\mathbb{C}$; we identify it with the group of invertible linear transformations of a one-dimensional complex vector space.

LEMMA 1.2. *Suppose* H *is a connected abelian Lie group. Write* L *for the kernel of the exponential map in* $\mathfrak{h}_0$, *and*

$$\Lambda_{\mathbb{C}} = \{\lambda \in \mathfrak{h}^* | \ \lambda(L) \subset 2\pi i\mathbb{Z}\}$$
$$\Lambda = \Lambda_{\mathbb{C}} \cap i(\mathfrak{h}_0)^*.$$

a) *Every irreducible unitary representation of* H *is of dimension one, and may therefore be regarded as a continous homomorphism from* H *into* $\mathbb{T}$. *Such a homomorphism is necessarily smooth.*

b) *Passage to differentials (that is, to the Lie algebra homomorphism attached to a group homomorphism) defines an identification of* Λ *with* $\hat{H}_u$. *If* λ *corresponds to* χ_{λ}, *then*

$$\chi_{\lambda}(\exp X) = \exp(\lambda(X)).$$

c) Write $\hat{H}$ for the set of irreducible one-dimensional representations of H. Then passage to differentials defines an identification of $\Lambda_{\mathbb{C}}$ with $\hat{H}$. In this way $\hat{H}_u$ is exhibited as a real manifold with $\hat{H}$ as a complexification.

Part (a) of this lemma is a consequence of spectral theory in Hilbert spaces (cf. [Weil, 1940]). The rest follows immediately from (for example) the identification of H with $\mathfrak{h}_0/L$ given by the exponential map.

We turn now to the specific structure theory for compact Lie groups that we will need.

Definition 1.3. Suppose K is a compact Lie group. A maximal torus in K is a maximal connected abelian subgroup T_0 of K.

A maximal torus T_0 is necessarily closed and therefore compact. Any two are conjugate under the group K. If K is the unitary group U(n) (of unitary operators on the Hilbert space $\mathbb{C}^n$), then T_0 may be chosen to be the group of diagonal unitary matrices. It is a product of n copies of U(1), which in turn is isomorphic to the circle group $\mathbb{T}$. Next, take K to be either the orthogonal group O(n) (consisting of real matrices in $U(n)$), or its identity component $SO(n)$ (consisting of elements of $O(n)$ of determinant 1). Then T_0 is a product of $[n/2]$ copies of $SO(2)$. The group $SO(2)$ consists of rotations of the plane $\mathbb{R}^2$, and may therefore also be identified with $\mathbb{T}$.

By a torus, we will mean in general a compact connected abelian Lie group. For such a group, the kernel L of the exponential map is a lattice in $\mathfrak{t}_0$; so Λ and $\Lambda_{\mathbb{C}}$ coincide. Any finite-dimensional representation of a torus T is a direct sum of irreducible representations. By Lemma 1.2, these may be identified with elements of Λ. If (π, V) is the representation, then we write the decomposition as

$$(1.4) \qquad\qquad V = \sum_{\lambda \in \Lambda} V_\lambda \ .$$

The subspaces V_λ are called *weight spaces*, and (1.4) is called the *weight space decomposition* of V. More explicitly,

$$(1.5) \qquad\qquad V_\lambda = \{v \in V \mid \pi(t)v = \chi_\lambda(t)v\}$$

(notation as in Lemma 1.2(b)). We write

$$(1.6) \qquad\qquad \Lambda(V) = \{\lambda \in \Lambda \mid V_\lambda \neq 0\} \ ,$$

the set of *weights* of V. It is often convenient to regard Λ as a multiset, with the multiplicity of λ equal to $\dim V_\lambda$.

Example 1.7. Suppose T is the group of diagonal matrices in $U(n)$. Using the identification of T with a product of n circles, we can identify the Lie algebra t_0 of T with $i\mathbb{R}^n$. (This may also be regarded as the n by n skew-hermitian diagonal matrices.) The kernel L of the exponential map is $\pi i\mathbb{Z}^n$. Identify t^* with $\mathbb{C}^n$ using the obvious pairing; then the lattice Λ of Lemma 1.2 is $\mathbb{Z}^n$. Consequently,

$$\hat{T} \cong \mathbb{Z}^n.$$

The identification works as follows. If $m = (m_1, \ldots, m_n)$ belongs to $\mathbb{Z}^n$, then

$$\chi_m\left[\text{diag}(z_1, \ldots, z_n)\right] = \prod_{i=1}^{n} (z_i)^{m_i}.$$

It is often easy to find the weight space decomposition of familiar representations of T. Suppose for example that V is the kth exterior power $\Lambda^k\mathbb{C}^n$. Then

$$\Delta(V) = \{(\delta_1, \ldots, \delta_n) \mid \delta_i \text{ is 0 or 1, and } \sum \delta_i = k\}.$$

Each weight has multiplicity one. The weight spaces are spanned by the standard basis vectors of the exterior algebra.

As a second example, consider the action of T by conjugation on n by n matrices. (This is the complexified adjoint representation (defined below) of $U(n)$, restricted

to T.) If we write $\{e_i\}$ for the usual basis of $\mathbb{Z}^n$, then
the weights of T are zero (with multiplicity n), and the
various $e_i - e_j$, for i not equal to j (with multiplicity
one). We leave to the reader the identification of the
weight spaces.

Any Lie group G acts on itself by conjugation. This
action has the identity as a fixed point, and so defines a
real representation (the *adjoint representation*) Ad of G
on the tangent space $\mathfrak{g}_0$ of G at e. The complexification of this is a representation of G on $\mathfrak{g}$, still denoted
Ad. In the case of a compact Lie group K, the restriction
of Ad to a maximal torus T_0 is a representation, to
which we can apply the discussion around (1.4). The non-
zero weights of T_0 are called *roots*; the set of all of
them is written

$$(1.8) \qquad\qquad \Delta(\mathfrak{k},\mathfrak{t}) \subset \hat{T} \subset i(\mathfrak{t}_0)^*.$$

(Notice the slight inconsistency with the notation (1.6),
which would suggest including zero in $\Delta(\mathfrak{k},\mathfrak{t})$.) Because of
the maximality of T_0, the zero weight space is precisely $\mathfrak{t}$.
We therefore have a *root space decomposition* of $\mathfrak{k}$:

$$(1.9) \qquad\qquad \mathfrak{k} = \mathfrak{t} \oplus \sum_{\alpha \in \Delta} \mathfrak{k}_\alpha .$$

Regarding the roots as linear functionals on t, we have

$$(1.10) \quad \mathfrak{k}_\alpha = \{X \in \mathfrak{k} \mid \text{ for all } H \in \mathfrak{t}, [H,X] = \alpha(H)X\}.$$

(This is a differentiated version of (1.5).) The root

spaces are all one-dimensional. It is clear from (1.10)

that complex conjugation on $\mathfrak{k}$ takes $\mathfrak{k}_\alpha$ to $\mathfrak{k}_{-\alpha}$. (This

uses the fact, discussed in the next paragraph, that the

roots take purely imaginary values on $\mathfrak{t}_0$.) The set of

roots is therefore closed under multiplication by -1.

Choose an element H^+ of $i\mathfrak{t}_0$ with the property that

$\alpha(H^+)$ is not zero for any α in $\Delta(\mathfrak{k},\mathfrak{t})$. (This is possible

since the set of roots is finite, and the kernel of each is

a proper subspace.) By the discussion preceding (1.4), all

weights take imaginary values on $\mathfrak{t}_0$. In particular, the

roots take real values on H^+. Define

(1.11) $\Delta^+(\mathfrak{k},\mathfrak{t}) = \{\alpha \in \Delta \mid \alpha(H^+) > 0\}$,

a set of positive roots for $\mathfrak{t}$ in $\mathfrak{k}$. Evidently

(1.12) $\Delta(\mathfrak{k},\mathfrak{t}) = \Delta^+(\mathfrak{k},\mathfrak{t}) \cup -\Delta^+(\mathfrak{k},\mathfrak{t})$,

a disjoint union.

PROPOSITION 1.13. *Suppose* K *is a compact Lie group and*

T_0 *is a maximal torus in* K. *Any two choices of a set of*

positive roots for $\mathfrak{t}$ *in* $\mathfrak{k}$ *are conjugate by the normali-*

zer of T_0 *in* K_0.

Definition 1.14. Suppose K is a compact Lie group, T_0

is a maximal torus in K, and Δ^+ is a set of positive roots for $\mathfrak{t}$ in $\mathfrak{k}$. The corresponding *Borel subalgebra* of $\mathfrak{k}$ is by definition

(a)
$$\mathfrak{b} = \mathfrak{t} \oplus \sum_{\alpha \in \Delta^+} \mathfrak{k}_\alpha .$$

The second summand on the right is the nil radical (that is, the largest nilpotent ideal) of $\mathfrak{b}$; it is denoted n. By the remarks after (1.10), the complex conjugate of n, which we denote n^-, is

(b)
$$n^- = \sum_{\alpha \in \Delta^+} \mathfrak{k}_{-\alpha}$$

Since $\mathfrak{t}$ is the complexification of a real subalgebra, it is equal to its complex conjugate. The complex conjugate of $\mathfrak{b}$ is therefore

(c)
$$\mathfrak{b}^- = \mathfrak{t} \oplus n^-$$

By (1.12),

(d)
$$\mathfrak{b} \cap \mathfrak{b}^- = \mathfrak{t}.$$

We define the *Cartan subgroup* T associated to T_0 and Δ^+ to be the normalizer in K of $\mathfrak{b}$:

(e)
$$T = \{t \in K \mid \mathrm{Ad}(t)\mathfrak{b} \subset \mathfrak{b}\}.$$

Later (Definition 1.28) we will generalize this definition, allowing certain subgroups between T and T_0. The special case defined by (e) will then be called a *large Cartan subgroup*.

It is easy to show that the Lie algebra $\mathfrak{b}$ is its own normalizer in $\mathfrak{k}$. Conseqently the Lie algebra of T is $\mathfrak{b} \cap \mathfrak{k}_0$, which equals $\mathfrak{t}_0$ by (d) in Definition 1.14. The identity component of T is therefore T_0. A more subtle point is that

$$(1.15) \qquad\qquad T \cap K_0 = T_0.$$

For connected compact Lie groups, a Cartan subgroup is therefore just a maximal torus.

Fix now a negative definite inner product $\langle \, , \, \rangle$ on $\mathfrak{k}_0$, invariant under $\mathrm{Ad}(K)$; this is possible since K is compact. The complexification of this inner product is a non-degenerate symmetric bilinear form on $\mathfrak{k}$, still denoted $\langle \, , \, \rangle$. Its restriction to $i\mathfrak{t}_0$ is positive definite, so by duality $i(\mathfrak{t}_0)^*$ acquires a positive definite inner product (which we also call $\langle \, , \, \rangle$).

Definition 1.16. In the setting of Definition 1.14, a weight λ in $\hat{T}_0$ is called dominant if

$$\langle \alpha, \lambda \rangle \geq 0,$$

for every root α in Δ^+. A representation (π, V) of T is called dominant if every weight occurring in its restriction to T_0 is dominant.

THEOREM 1.17 (Cartan-Weyl). *Suppose* K *is a compact Lie group, and* T *is a Cartan subgroup (Definition 1.14). There is a bijection between the set* $\hat{K}_u$ *of irreducible unitary representations of* K *and the set of irreducible dominant unitary representations of* T *(Definition 1.16), defined as follows. If* (π,V) *is an irreducible unitary representation of* K, *define* V^+ *to be the subspace annihilated by* n *(Definition 1.14). Then* V^+ *is invariant under* T, *and the corresponding representation* π^+ *on* V^+ *is irreducible and dominant.*

Let us consider the extent to which this "computes" $\hat{K}_u$. If K is connected, then Theorem 1.17 says that the irreducible representations of K are parametrized by the dominant weights. This is a completely computable and satisfactory parametrization (cf. Example 1.18 below), although of course one can ask much more about how the representation is related to the weight. If K is finite, then T is equal to K, and the theorem is a tautology; it provides no information about $\hat{K}_u$. In general, if K is not connected then T is not connected either, and its representation theory can be difficult to describe explicitly. Essentially the problem is one about finite groups, and we should not be unhappy to treat it separately.

Example 1.18. Suppose K is $U(n)$, $T = T_0$ is the group of diagonal matrices, and

$$\Delta^+ = \{e_i - e_j \mid i < j\}$$

(cf. Example 1.8). If we identify t^* with $\mathbb{C}^n$, we may take $\langle \, , \, \rangle$ to be the standard quadratic form. The set of dominant weights is then

$$\{\lambda = (\lambda_1, \ldots, \lambda_n) \mid \lambda_i \in \mathbb{Z}, \text{ and } \lambda_i \geq \lambda_j \text{ if } i < j\},$$

the set of non-increasing sequences of n integers.

We turn now to the problem of realizing the representations described in Theorem 1.17. To do so requires constructing a certain complex manifold on which K acts transitively. Here is a general recipe for doing that.

PROPOSITION 1.19. *Suppose* G *is a Lie group and* H *is a closed subgroup. The set of* G-*invariant complex structures on* G/H *is in natural bijection with the set of subalgebras* $\mathfrak{b}$ *of* $\mathfrak{g}$ *(the complexified Lie algebra of* $\mathfrak{g}$*), having the following properties:*

a) $\mathfrak{b}$ *contains* $\mathfrak{h}$, *and* $\mathrm{Ad}(H)$ *preserves* $\mathfrak{b}$;

b) *the intersection of* $\mathfrak{b}$ *with its complex conjugate* $\mathfrak{b}^-$ *is precisely* $\mathfrak{h}$; *and*

c) the dimension of $\mathfrak{b}/\mathfrak{h}$ *is half the dimension of*
$\mathfrak{g}/\mathfrak{h}$. *In the usual identification of* $\mathfrak{g}/\mathfrak{h}$ *with the com-*
plexified tangent space of G/H *at* eH, *the subspaces* $\mathfrak{b}/\mathfrak{h}$
and $\mathfrak{b}^-/\mathfrak{h}$ *correspond to the holomorphic and antiholomorphic*
tangent spaces, respectively.

Proposition 1.21 will explain how to describe the holomorphic functions on G/H in terms of $\mathfrak{b}$.

PROPOSITION 1.20. *Suppose* G *is a Lie group and* H *is a*
closed subgroup. The set of C^∞ *homogeneous vector bundles*
on G/H *is in natural bijection with the set of finite-*
dimensional representations of H, *by sending a vector*
bundle $\mathscr{W}$ *to its fiber* W *over* eH. *Using this bijection,*
we may identify the space $C^\infty(G/H, \mathscr{W})$ *of smooth sections of*
$\mathscr{W}$ *with the space of* W-*valued smooth functions* f *on* G,
satisfying

   a) $\qquad\qquad f(gh) = \tau(h^{-1})f(g).$

Here of course τ *denotes the isotropy action of* H *on* W.

A proof of this result will be sketched in Chapter 3 (Proposition 3.2 and Corollary 3.4).

PROPOSITION 1.21. *Suppose G/H carries an invariant*
complex structure given by $\mathfrak{b}$ (Proposition 1.19) and a
homogeneous vector bundle $\mathscr{W}$ given by W (Proposition
1.20). Then to make $\mathscr{W}$ a holomorphic vector bundle amounts
to giving a Lie algebra representation (also called τ) of
$\mathfrak{b}^-$ on W, satisfying

   a) the differential of the group representation of H
agrees with the Lie algebra representation restricted to $\mathfrak{h}$;
and

   b) for h in H, X in $\mathfrak{b}^-$, and w in W,
$$\tau(h)[\tau(X)w] = \tau(Ad(h)X)[\tau(h)v] \quad .$$
The space $\Gamma(G/H,\mathscr{W})$ of holomorphic sections of $\mathscr{W}$ may be
identified with the space of W-valued smooth functions f
on G, satisfying Proposition 1.20(a) and the following con-
dition. For every X in $\mathfrak{b}^-$, we require that
$$(Xf)(g) = -\tau(X)(f(g)).$$
Here the action on the left comes from regarding the Lie
algebra as left-invariant vector fields on G.

To set up the Borel-Weil theorem, we need just one more
observation.

LEMMA 1.22. *In the setting of Theorem 1.17, the orthogonal*
complement of V^+ in V is

$$V^o = \pi(n^-)V.$$

Consequently, V^+ may be identified (as a representation of T) with the quotient V/V^o.

Proof. Write σ for the complex conjugation automorphism of $\mathfrak{k}$:

(1.23) $\sigma(X + iY) = X - iY$ (X,Y in $\mathfrak{k}$).

The inner product $\langle \, , \, \rangle$ on V satisfies

$$\langle \pi(Z)v_1, v_2 \rangle = -\langle v_1, \pi(\sigma Z)v_2 \rangle \quad .$$

(Here as usual we write π for the differentiated representation of the Lie algebra.) Since $\sigma(n^-)$ is n, an easy formal argument now shows that the orthogonal complement of $\pi(n^-)V$ is V^+. The assertion of the lemma follows.   □

THEOREM 1.24 (Borel-Weil). Suppose K is a compact Lie group, T is a Cartan subgroup, and $\mathfrak{b}$ is a Borel subalgebra normalized by T (Definition 1.14). Fix an irreducible representation (τ, W) of T, and extend the differentiated representation of $\mathfrak{t}$ to $\mathfrak{b}^-$ by making n^- act trivially. Write $\mathscr{W}$ for the resulting holomorphic vector bundle on K/T (Propositions 1.19, 1.20, and 1.21). Let

$$V = \Gamma(K/T, \mathscr{W}),$$

K act on V by left translation of sections; the resulting representation is denoted π. Then (π,V) is non-zero if and only if (τ,W) is a dominant representation of T. In that case, (π,V) is the irreducible representation of K attached to (τ,W) by Theorem 1.17. That is, (τ,W) is naturally isomorphic to the representation (π^+,V^+) of T on the subspace of V annihilated by n.

Example 1.25. Suppose K is U(2), and T is the group of diagonal matrices. K acts transitively on the projective space $\mathbb{CP}^1$ of complex lines in $\mathbb{C}^2$. The stabilizer of the line through the second coordinate is T; so K/T may be identified with $\mathbb{CP}^1$. This defines the complex structure.

A typical homogeneous line bundle is the tautological bundle, which puts over every point of $\mathbb{CP}^1$ the line which it "is." In the identification of homogeneous bundles with characters of T, which are in turn identified with $\mathbb{Z}^2$, the tautological bundle corresponds to $(0,1)$. That weight is not dominant (cf. 1.18); so the Borel-Weil theorem says that the bundle should have no sections. This well-known fact simply means that the only holomorphic way to pick a point in each line in $\mathbb{C}^2$ is to pick zero everywhere.

The dual line bundle to the tautological bundle asso-
ciates to each line the space of linear functionals on it;
this corresponds to the weight $(0,-1)$, which is dominant.
We can find global sections of the bundle by fixing a linear
functional on $\mathbb{C}^2$ and restricting it to each of the lines.
All the holomorphic sections arise in this way; so the space
of sections forms a two-dimensional representation of K.

We are going to prove the Borel-Weil theorem; more precisely, we will show how to deduce it from Theorem 1.17.   It
is helpful to introduce a definition.

Definition 1.26.   Suppose H is a Lie group, and $\mathfrak{b}$ is a
complex Lie algebra.   Assume that we are given

    i) an inclusion of the complexified Lie algebra $\mathfrak{h}$ of
H into $\mathfrak{b}$; and

    ii) an action, denoted Ad, of H on $\mathfrak{b}$ by automor-
phisms, extending the adjoint action on $\mathfrak{h}$.
A $(\mathfrak{b},H)$-*module* is a complex vector space V (possibly
infinite-dimensional), carrying a group representation of H
and a Lie algebra representation of $\mathfrak{b}$, subject to the following three conditions.   (For the moment we will write π
for these representations, but later it will usually be

conenient to drop this in favor of module notation: h·v
instead of π(h)v.)

a)   The group representation is locally finite and
smooth.   That is, if   v ∈ V, the vector space span of π(H)v
is finite-dimensional, and   H   acts smoothly (or, equivalently, continuously) on this space.

b)   The differential of the group representation (which
exists by (a)) is the restriction to $\mathfrak{h}$ of the Lie algebra
representation.

c)   The group representation and Lie algebra representation are compatible in the sense that

$$\pi(h)\pi(X) = \pi(Ad(h)X)\pi(h)$$

For elements h in H_0, condition (c) follows from (b).

If   V   and   W   are   ($\mathfrak{b}$,H)-modules, we can form
$\text{Hom}_{\mathfrak{b},H}(V,W)$.   This is the space of linear transformations
from   V   to   W   that are compatible with both representations.

The next result is a version of Frobenius reciprocity
appropriate for the Borel-Weil theorem.

PROPOSITION 1.27.   *Suppose   G   is a Lie group,   H   is a*
closed subgroup, $\mathfrak{b}$ defines a holomorphic structure on   G/H
(Proposition 1.19), and   W   is a ($\mathfrak{b}^-$,H)-module corresponding

to a holomorphic vector bundle $\mathscr{W}$ on G/H (Proposition

1.21). Let V be a finite-dimensional representation of

G; by differentiation and restriction, we may regard V as

a $(\mathfrak{b}^-,H)$-module. Then there is a natural isomorphism

$$\text{Hom}_G(V,\Gamma(G/H),\mathscr{W}) \cong \text{Hom}_{\mathfrak{b}^-,H}(V,W) \ .$$

Proof. Write Φ for a typical element on the left, and ϕ

for a typical element on the right. If these correspond

under the isomorphism, then

$$\phi(v) = [\Phi(v)](e)$$
$$[\Phi(v)](g) = \phi(\pi(g^{-1})v).$$

(We are using the description of $\Gamma(G/H,\mathscr{W})$ contained in

Proposition 1.21.) The verification that these formulas

define the isomorphism we want is left to the reader.   □

Proof of Theorem 1.24. Let (ξ,X) be any finite-

dimensional representation of K. By Proposition 1.27,

$$\text{Hom}_G(X,V) \cong \text{Hom}_{\mathfrak{b}^-,T}(X,W).$$

Now n^- acts trivially on W; so the right side is

$$\text{Hom}_T(X/\xi(n^-)X],W)$$

By Lemma 1.22, we therefore have

$$\text{Hom}_G(X,V) \cong \text{Hom}_T(X^+,W).$$

By Theorem 1.17, the right side is always zero unless W is dominant; and in that case it is one-dimensional for a unique irreducible representation X.   □.

We will conclude this chapter with two reformulations of Theorem 1.17, each of which will be convenient or instructive later on.

Definition 1.28. Suppose K is a compact Lie group. Use the notation of Definition 1.14 (a)-(d). Write T^+ for the large Cartan subgroup of Definition 1.14 (e). The *small Cartan subgroup* associated to T_0 is

(a) $T^- = \{t \in K | \text{Ad}(t) \text{ is trivial on } \mathfrak{t}\}$.

A *general Cartan subgroup* associated to T_0 and Δ^+ is a subgroup T between T^- and T^+.

Suppose T is a small Cartan subgroup of K. Write $N_K(T)$ for the normalizer of T in K. The *Weyl group* of T in K is the quotient

(b) $W(K,T) = N_K(T)/T$.

If $\text{Ad}(K)$ consists of inner automorphisms (that is, if K is of Harish-Chandra's class in the sense of Definition

0.6), then the notions of small and large Cartan subgroup coincide.

For the next definition, we need to make sense of the *length* of a weight. This is defined using the inner product $\langle \, , \, \rangle$ on $i(\mathfrak{t}_0)^*$ (defined before (1.16).

Definition 1.29. Suppose T is a Cartan subgroup of K, and (π,V) is a finite-dimensional representation of K. An irreducible representation τ of T is called *extremal* in π if

a) τ occurs in the restriction of π to T; and

b) the length of any weight λ of τ (cf. (1.4) is greater than or equal to the length of any weight of π.

Here is the first reformulation of Theorem 1.17.

THEOREM 1.30 (*Cartan-Weyl*). *Suppose* K *is a compact Lie group. Write* $\Phi(K)$ *for the set of* K-*conjugacy classes of pairs* (T,τ), *with* T *a small Cartan subgroup of* K *and* τ *in* $\hat{T}$.

a) Fix a particular small Cartan subgroup T, *with Weyl group* W. *Then* $\Phi(K)$ *may be identified with* $\hat{T}/W$, *the set of* W *orbits on* $\hat{T}$.

b) *There is a finite-to-one correspondence from* $\hat{K}$
onto $\Phi(K)$, *defined by associating to* π *the set of extremal*
representations of small Cartan subgroups occurring in π.

c) *If* K *is of Harish-Chandra's class (Definition*
0.6), then the correspondence in (b) is a bijection.

We will not prove this result in detail. Note, however,
that (a) is a formal consequence of the conjugacy of all
maximal tori in K. The other fact needed in the proof (or
rather the reduction to Theorem 1.17) is that every representation of T is conjugate under W to a dominant one.
This in turn follows from Proposition 1.13.

The second reformulation of Theorem 1.17 is motivated
by the theory of characters (Definition 1.38 below). It
turns out that characters of compact groups are most naturally expressed as quotients of two multi-valued functions
(Theorem 1.40). These functions will be single-valued on
certain coverings of Cartan subgroups. In order to define
these coverings, we recall a standard general construction.

Definition 1.31. Suppose $H^{\sim}$ is a topological group, F
is a closed normal subgroup, and H is the quotient group
$H^{\sim}/F$. Let G be another topological group, and τ a

homomorphism from G to H. Define a new group $G^\sim$, the
pullback from H *to* G *of* $H^\sim$, by

$$G^\sim = \{(x,g) \in (H^\sim \times G) \mid \pi(x) = \tau(g)\}.$$

Here π denotes the quotient map from $H^\sim$ to H. Then $G^\sim$
contains a copy of F (as $F \times \{e\}$). Projection on the
second factor defines a surjection from $G^\sim$ to G, with
kernel F. Finally, projection on the first factor gives a
map $\tau^\sim$ from $G^\sim$ to $H^\sim$. We therefore have a commutative
diagram

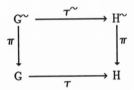

An important case is the extraction of nth roots of characters. In that case, we take H to be the multiplicative
group $\mathbb{C}^\times$ of non-zero complex numbers. $H^\sim$ is again $\mathbb{C}^\times$,
and the map π is

$$\pi(z) = z^n.$$

The group F is the group of nth roots of unity in $\mathbb{C}^\times$,
which is isomorphic to $\mathbb{Z}/n\mathbb{Z}$. We take G arbitrary, and τ
any character of G. Then $\tau^\sim$ is an nth root of τ (in
the sense that $n\tau^\sim$ descends from $G^\sim$ to G and coincides
with τ there).

Here is a useful condition for the equivalence of two pullback covers.

LEMMA 1.32. *Suppose* G *is a topological group, and* τ_1 *and* τ_2 *are characters of* G. *Assume that there is a third character* ϕ *of* G, *with the property that (with multiplication of characters written additively)*

a) $\tau_2 = \tau_1 + n\phi$.

Then there is an isomorphism (depending on ϕ) *between the coverings of* G *induced by* nth *roots of* τ_1 *and* τ_2.

  Conversely, any isomorphism of the coverings compatible with the projections to G *and the inclusions of* $\mathbb{Z}/n\mathbb{Z}$ *arises from a unique character* ϕ *satisfying* (a).

We leave this as an exercise for the reader.

  Here are the coverings we need.

Definition 1.33. Fix notation as in Definition 1.28; in particular, T is a Cartan subgroup of K. Write 2ρ for the determinant character of T on n:

$$2\rho(t) = \det(\mathrm{Ad}(t)|_n).$$

The differential of 2ρ is denoted by the same symbol; it is the sum of the positive roots of $\mathfrak{k}$. The two-fold cover

$T^\sim$ of T, defined by the square root ρ of 2ρ, is called

the *metaplectic cover* of T (for reasons which will become

clearer in Chapter 10 - see Proposition 10.17). The *Weyl*

denominator is the function

$$\Delta(t^\sim) = \rho((t^\sim)^{-1})[\det (I - Ad(t)|_n)]$$

on $T^\sim$; here of course t denotes the image of $t^\sim$ in T.

Write ζ for the non-trivial element of the kernel of

the covering. Then $\rho(\zeta) = -1$; so

(1.34)(a) $\Delta(\zeta x) = -\Delta(x)$.

Suppose $X \in t^o$; write $\exp^\sim(X)$ for the exponential of X

in $T^\sim$. Then an easy calculation shows that

(1.34)(b) $\Delta(\exp^\sim(X)) = \prod_{\alpha \in \Delta^+} (e^{\alpha(X)/2} - e^{-\alpha(X)/2})$.

It may happen that T has a character with differen-

tial ρ. In that case G is called *acceptable*. When G

is acceptable, Lemma 1.32 implies that the covering $T^\sim$ is

isomorphic to $T\times(\mathbb{Z}/2\mathbb{Z})$, and the results below can be formu-

lated on T itself. For noncompact G, Harish-Chandra uses

an analogous assumption in much of his work on the discrete

series. If G is connected, it turns out that either G

or some double cover of G is acceptable; so there is

little loss of generality in making the assumption. Unfortu-

nately, the more general coverings of Definition 5.7 cannot

be trivialized by passing to a covering of G. We will

therefore keep track of coverings here as well, as practice for the general case.

PROPOSITION 1.35 (the Weyl denominator formula). *In the setting of Definitions 1.28 and 1.33, suppose* $t^\sim$ *belongs to the preimage* $(T^-)^\sim$ *of the small Cartan subgroup in* $T^\sim$. *Then*

$$\Delta(t^\sim) = \sum_{w \in W_0} \epsilon(w) w\rho(t^\sim).$$

Here W_0 *denotes the Weyl group of* T^- *in* K_0, *and* $\epsilon(w)$ *is the determinant of the action of* w *on* $\mathfrak{t}_0$.

This proposition is a special case of the Weyl character formula, to which we will turn in a moment. First, however, we should explain what is meant by the character $w\rho$ of $(T^-)^\sim$ that appears in the proposition. Write $[(T^-)^\sim]'$ for the covering defined by the square root of $w(2\rho)$. This has a well-defined character $w\rho$. To define $w\rho$ on $(T^-)^\sim$, we need to define an isomorphism between the two coverings. It can be shown that $w(2\rho)-2\rho$ is of the form 2ϕ, with ϕ a sum of roots. Any two sums of roots with the same differential define the same character of T^-; so ϕ is uniquely defined. Lemma 1.32 now provides the isomorphism we need.

Definition 1.36. Suppose T is a (large) Cartan subgroup of the compact group K, associated to the Borel subalgebra $\mathfrak{b}$. Let $T^\sim$ be the metaplectic cover of T (Definition 1.33), and ζ the nontrivial element of the kernel of the covering map. A *dominant regular metaplectic representation* of T is a representation τ of $T^\sim$, with the following properties:

a) $\tau^\sim(\zeta) = -\text{Id}$; and

b) if λ is any weight of $(T_0)^\sim$ occurring in $\tau^\sim$, then λ is dominant and regular. That is,

$$\langle \alpha, \lambda \rangle > 0,$$

for every positive root α of $\mathfrak{t}$ in $\mathfrak{k}$.

Here is the second reformulation of Theorem 1.17.

THEOREM 1.37 (Cartan-Weyl). *Suppose K is a compact Lie group, and T is a (large) Cartan subgroup. Then there is a bijection between the set $\hat{K}_u$ of irreducible unitary representations of K, and the set of irreducible dominant regular metaplectic representations of $T^\sim$ (Definition 1.36).*

Proof. Notice first that representations of T are the same as representations of $T^\sim$ taking ζ to 1. With this

identification, it is an easy exercise to check that tensoring with the character ρ defines a bijection from the parameter set in Theorem 1.17 to that in Theorem 1.37.   □

It is possible to make a variety of other reformulations of Theorem 1.17 by combining the ideas in Theorems
1.30 and 1.37, or by including a Mackey-type analysis of the
representations of large Cartan subgroups. We leave this to
the imagination of the reader.

This proof of Theorem 1.37 offers no hint of why one
ought to introduce T~. Its usefulness first appears in the
theory of characters.

Definition 1.38. Suppose (π, V) is a finite-dimensional
representation of the group G. The *character of* π is the
(continuous) function

$$\theta_\pi(g) = \text{tr } \rho(g)$$

on G. Notice that θ is a class function: that is, it is
constant on conjugacy classes in G.

The character plays an important part in harmonic analysis,
particularly in connection with the Plancherel formula for
G. One piece of information that it obviously contains is
the dimension of π, which is $\theta_\pi(1)$. Rather than discuss-

ing anything more sophisticated, we will content ourselves
with justifying characters by the following easy result.

PROPOSITION 1.39. *Two finite-dimensional irreducible representations of a group G have the same character if and
only if they are equivalent. More generally, the characters
of any family of inequivalent irreducible representations
are linearly independent (as functions on G).*

To prove this, one introduces the convolution algebra of
functions on G with finite support. In this way one is
reduced to the case of representations of algebras. There
the proposition follows immediately from the Wedderburn
theorem.

We would like to compute the characters of compact Lie
groups as explicitly as possible. Because they are class
functions, it is enough to compute them on some subset of G
that meets every conjugacy class. A (large) Cartan subgroup
has this property.

THEOREM 1.40 (the Weyl character formula: first version).
*Suppose K is a compact Lie group, and T is a Cartan sub-
group attached to the Borel subalgebra 𝔟. Let π be an
irreducible representation of K, and τ~ the corresponding*

irreducible dominant regular metaplectic representation of $T^\sim$. *Let* T^- *be the small Cartan subgroup inside* T *(Definition 1.28). Write* $(\tau^-)^\sim$ *for the restriction of* $\tau^\sim$ *to* $(T^-)^\sim$. *Then for any* $t^\sim$ *in* $(T^-)^\sim$, *with image* t *in* T^-, *we have*

$$\Delta(t^\sim)\theta_\pi(t) = \sum_{w\in W_0} \epsilon(w)\ w(\tau^-)^\sim(t^\sim).$$

Here Δ is the Weyl denominator (Definition 1.33); and $\epsilon(w)$ is the determinant of the action of w on $\mathfrak{t}_0$. The characters $w(\tau^-)^\sim$ of $(T^-)^\sim$ are defined just as in the remarks after Proposition 1.35.

Weyl's original proof of this theorem exploited the connection of characters with the Plancherel formula. He showed that there was a formula of approximately this form (which follows from Theorem 1.30) and then used formal facts about harmonic analysis on G to deduce the precise form. Most of the known proofs have the same general shape, although they may exploit different "formal facts." A nice algebraic account is in [Humphreys, 1972].

The Weyl denominator vanishes only on a lower dimensional set in $T^\sim$, so knowledge of $\Delta\theta_\pi$ on $T^\sim$ determines θ_π almost everywhere. For many analytic purposes, this is good enough. If $\mathfrak{t}$ is not abelian, however, the identity element is one of the excluded points; so Theorem 1.40 as it

stands does not compute the dimension of π (see the remark
after Definition 1.38). This defect may be remedied by
applying L'Hopital's rule to evaluate the limit as t approaches 1. Here is the result.

COROLLARY 1.41 (the Weyl dimension formula). *In the setting
of Theorem 1.40, fix a weight* λ *in* t^* *of* $\tau^{\sim}$. *Then*

$$\dim \pi = (\dim \tau^{\sim}) \prod_{\alpha \in \Delta^+} \langle \alpha, \lambda \rangle / \langle \alpha, \rho \rangle. \quad .$$

The most serious problem with Theorem 1.40 is that the
formula it gives is valid only on the small Cartan subgroup
T^-. For K in Harish-Chandra's class (Definition 0.6), T^-
coincides with T, and there is no problem. This is certainly the most important case; but for completeness, we
will sketch the extension of Weyl's formula to all of T.

Notice first that T normalizes t_0. By Definition
1.28(a), it follows that T normalizes T^-, and therefore
also its normalizer in K_0. By Definition 1.28(b), this
means that T acts on the Weyl group W_0 of T^- in K_0.
For w in W_0, define

$$(1.42) \qquad T_w = \{t \in T \mid t \cdot w = w\}.$$

An element w of W_0 is determined by the image $w\flat$ of $\flat$
under w. Since T normalizes $\flat$, on checks easily that

$$(1.42)' \qquad T_w = \{t \in T \mid t \text{ normalizes } w\flat\}.$$

The characters 2ρ and $2w\rho$ of T_w therefore both make sense. In addition, their difference has a natural square root: the determinant of the action of T_w on

$$\mathfrak{b}/(\mathfrak{b} \cap \mathfrak{b}_w).$$

The reasoning after Proposition 1.35 now provides an action of w on $[(T_w)^\sim]^\wedge$.

THEOREM 1.43 (the Weyl character formula: second version). *In the setting of Theorem 1.40, suppose $t^\sim$ in $T^\sim$ has image t in T. Put*

$$W_0(t) = \{w \in W_0 \mid t \in T_w\}.$$

Then

$$\Delta(t^\sim)\theta_\pi(t) = \sum_{w \in W_0(t)} \epsilon(w) \ w(\tau^\sim)(t^\sim)$$

Here $w\tau^\sim$ is defined only on $(T_w)^\sim$, in accordance with the remarks preceding the proposition.

Most of the proofs of Theorem 1.40 may be adapted to yield Theorem 1.43. This is true in particular of Weyl's original proof.

Chapter 2

HARISH-CHANDRA MODULES

In this chapter, we will present the ideas developed by
Harish-Chandra (mostly in [Harish-Chandra, 1953]) for reduc-
ing the infinite-dimensional representation theory of reductive groups to algebra. To begin, we need a little structure theory. Convenient references are [Knapp, 1986] or
[Warner, 1972].

Let G be a reductive group (Definition 0.6). Fix
once and for all a maximal compact subgroup K of G. Let
s_0 be an Ad(K)-invariant complement for $\mathfrak{k}_0$ in $\mathfrak{g}_0$. (It
is much more usual to call this space $\mathfrak{p}_0$, but we prefer to
reserve $\mathfrak{p}_0$ for parabolic subalgebras.) Then G is diffeo-
morphic to the product of K and s_0 under the obvious map:

(2.1) $G = K \cdot \exp(s_0)$.

This suggests that all the essential obstructions to passing
from the Lie algebra to the Lie group should involve only

50

K. Harish-Chandra's results make this precise for representation theory.

Choose an Ad(G)-invariant symmetric bilinear form
$\langle \ , \ \rangle$ on $\mathfrak{g}_0$, positive definite on $\mathfrak{s}_0$ and negative definite on $\mathfrak{k}_0$, and making these two subspaces orthogonal. We
will use the same notation for various restrictions and
complexifications of the form.

By (2.1), we can define a map θ from G to G, by

$$(2.2) \qquad \theta(k \cdot \exp(X)) = k \cdot \exp(-X).$$

It turns out that θ is an automorphism of G, the *Cartan
involution*. We use the same letter for its differential, an
involutive automorphism of $\mathfrak{g}_0$.

Definition 2.3. Suppose (π, V) is a representation of a
Lie group. The space of *smooth vectors* V^∞ in V consists
of those v such that the map

$$g \longrightarrow \pi(g)v$$

from G to V is smooth.

Obviously V^∞ is invariant under the action of G. If V
is a reasonable complete space, such as a Frechet space,
then V^∞ is dense in V. (To prove this, one needs only to
be able to integrate compactly supported smooth functions

with values in V.) V^∞ can be given a natural topology, so
that π defines a representation π^∞ of G on V^∞. This
representation is smooth, and may therefore be differentiated to give a Lie algebra representation of $\mathfrak{g}$.

Definition 2.4. Suppose (π,V) is a representation of a
topological group G, and K is a compact subgroup of G.
The space V_K of K-<u>finite vectors</u> in V consists of those
v such that the set $\pi(K)v$ spans a finite-dimensional subspace of V.

The subspace V_K is dense in V for reasonable complete
spaces V; again the integrability of compactly supported
smooth V-valued functions suffices. However, V_K is not
in general invariant under G. Harish-Chandra's circumven-
tion of this problem begins with the following easy fact.

PROPOSITION 2.5. *Suppose* (π,V) *is a representation of a
Lie group* G *and* K *is a compact subgroup of* G. *Then the
space* $X = (V^\infty)_K$ *of K-finite smooth vectors in* V *is
invariant under the representation* π^∞ *of* $\mathfrak{g}$. *In this way*
X *acquires the structure of a* $(\mathfrak{g},K)$-*module* (Definition
1.26).

The $(\mathfrak{g}, K)$-module X is called the *Harish-Chandra module* of π. Notice that X has no topology; everything we want to say about such modules will be essentially algebraic in nature.

Example 2.6. Let G be $SU(1,1)$, the group of 2 by 2 complex matrices of determinant one, preserving the form $|z_1|^2 - |z_2|^2$ on $\mathbb{C}^2$. One checks easily that G consists of all matrices

$$g(\alpha, \beta) = \begin{bmatrix} \alpha & \beta \\ \bar{\beta} & \bar{\alpha} \end{bmatrix}$$

such that

$$|\alpha|^2 - |\beta|^2 = 1.$$

The group G acts on the unit circle $\mathbb{T}$, by linear fractional transformations:

$$g(\alpha, \beta)^{-1} \cdot z = (\alpha z + \beta)/(\bar{\beta} z + \bar{\alpha}) .$$

This gives rise to a representation π of G on $V = L^2(\mathbb{T})$, by

$$[\pi(g)f](z) = f(g^{-1} \cdot z).$$

The operators $\pi(g)$ are not unitary, because the action does not preserve the measure. They are bounded, however. We have

$$V^\infty = C^\infty(\mathbb{T})$$

$$V_K = \text{trigonometric polynomials on } \mathbb{T}.$$

It is a good exercise to compute explicitly the action of

the Lie algebra on the trigonometric polynomials, and verify

Theorem 2.10 below in this case. (This is essentially

carried out in section 1.1 of [Vogan, 1981], for example.)

There are exactly three proper closed invariant subspaces of

V: the constants; the boundary values of functions holomor-

phic in the disc; and the boundary values of the functions

holomorphic in the Riemann sphere minus the disc.

Our discussion of K-finite vectors will be clarified

by some general facts about representations of compact

groups.

PROPOSITION 2.7. *Suppose* K *is a compact topological*

group.

   a) *Suppose* (π,V) *is an irreducible representation of*

K. *If the space* V *admits at least one non-zero continuous*

linear functional, then V *is finite-dimensional.*

   b) *Suppose* (π,V) *is a finite-dimensional*

representation of G. *Then there is an inner product* $\langle\ ,\ \rangle$

on V *making* π *a unitary representation. If* π *is irre-*

ducible, then $\langle\ ,\ \rangle$ *is unique up to a scalar multiple.*

Because of part (a) of this proposition, it is reasonable to define $\hat{K}$ to be the set of equivalence classes of irreducible finite-dimensional representations of K (cf. Definition 0.4 and Lemma 1.2). Then $\hat{K}$ is the same set as $\hat{K}_u$, but we will use the former notation when the unitary structure is not given to us.

As a consequence of Proposition 2.7(b), any K-finite representation V may be decomposed as

$$(2.8) \qquad V = \sum_{\delta \in \hat{K}} V_\delta.$$

Here V_δ is a sum of copies of δ; the number of copies is a well-defined cardinal number $m(\delta, V)$, *the multiplicity of δ in* V. By Schur's lemma,

$$2.9) \qquad m(\delta, V) = \dim \mathrm{Hom}_K(Z_\delta, V);$$

here Z_δ is the space for a copy of δ. We can use (2.10) to define the multiplicity of δ in any representation V (not necessarily K-finite); this amounts to considering the multiplicity in the space of K-finite vectors. Similarly, the subspace V_δ is defined in general, and we have

$$(2.10) \qquad V_K = \sum_{\delta \in \hat{K}} V_\delta.$$

The subspace V_δ is called the δ-primary or δ-isotypic subspace.

Definition 2.11. A representation (π,V) of a compact group K is called admissible if each irreducible representation of K has finite multiplicity in V.

For example, any reasonable space of functions (continuous, L^p, generalized, etc.) on a homogeneous space for K will be admissible by Frobenius reciprocity: the multiplicity of δ will be the dimension of the subspace of Z_δ fixed by an isotropy group. This is proved in the same way as Proposition 1.27.

Here is the first serious result of the chapter.

THEOREM 2.12 ([Harish-Chandra, 1953]). Suppose G is a Lie group, and K is a compact subgroup meeting every component of G. Let (π,V) be an admissible representation of G on a Banach space.

a) K-finite vectors in V are automatically smooth: $V_K \subset V^\infty$. In particular, the space X of K-finite vectors is a (g,K)-module.

b) *There is a bijection between* (g,K)*-invariant sub-*
spaces of X, *and closed* G-*invariant subspaces of* V. *It*
is defined in one direction by passage to K-*finite vectors*
and in the other direction by passage to closure.

Part (a) here is a consequence of the remark after Definition 2.4. For (b), one has to check that the closure of a
(g,K)-submodule is G-invariant. To see why there is some-
thing to prove, consider the representation of $\mathbb{R}$ on $L^2(\mathbb{R})$
by translation. The smooth vectors are smooth functions
with all derivatives in L^2, and the action of the Lie algebra is by differentiation. The subspace of smooth functions
supported on $[-1,1]$ is invariant under differentiation;
but its closure (which is $L^2([-1,1])$) is not translation
invariant. What is needed is a notion of analytic vector,
and the existence of many of them. This was proved by
Harish-Chandra in the setting of the theorem (and in great
generality in [Nelson, 1959]). The main point is that
K-finite vectors turn out to be analytic (when π is
admissible).

The next point to check is that the definition of
admissible does not exclude all the interesting representa-

ions.   (For non-reductive groups, it usually does exactly
that).

THEOREM 2.13 (Harish-Chandra).   *Suppose* G *is a reductive*
Lie group and K *is a maximal compact subgroup.*

   a) *Any irreducible unitary representation of* G *is*
admissible.

   b) *Suppose* π_1 *and* π_2 *are irreducible unitary*
representations of G, *and that the Harish-Chandra modules*
of π_1 *and* π_2 *(defined after Proposition 2.5) are isomor-*
phic as (g,K)-*modules. Then* π_1 *and* π_2 *are unitarily*
equivalent.

The idea of the proof of (a) is this.   It is not difficult
to reduce to the case when G is connected.   Then Segal had
shown that the unitarity assumption makes available enough
spectral theory to force the center $Z(g)$ of $U(g)$ to act
by scalars on V^∞.   Roughly speaking, Harish-Chandra showed
that certain large pieces of the enveloping algebra are in
some sense integral over $Z(g)$, and must therefore act in a
locally finite way.   (This is the most difficult step.)
Admissible (g,K)-submodules of V^∞ therefore exist.   Using
the theory of analytic vectors, he deduced (a).

Definition 2.14. Suppose G is a reductive group, and K
is a maximal compact subgroup. Two admissible representations of G are said to be *infinitesimally equivalent* if
their Harish–Chandra modules are isomorphic as $(\mathfrak{g},K)$–
modules. We write $\hat{G}$ for the set of infinitesimal equivalence classes of irreducible admissible representations of
G; this set contains $\hat{G}$ by Theorem 2.13.

The reader may wish to compare our definitions of $\hat{G}$ in the
compact and abelian cases (Lemma 1.2 and after Proposition
2.6) and note that Definition 2.14 is consistent with those.

   Theorem 2.13 allows one to recover at least unitary
representations from their attached $(\mathfrak{g},K)$–modules. It is
natural to ask whether *any* $(\mathfrak{g},K)$–module arises from a representation. This is not easy to prove, but at least for irreducibles it is true.

THEOREM 2.15 ([Lepowsky, 1973]). *Suppose G is a reductive
Lie group and K is a maximal compact subgroup. Let X be
an irreducible $(\mathfrak{g},K)$-module (Definition 1.26). Then*

   a) X *is admissible (Definition 2.11); and*

   b) X *is isomorphic to the Harish-Chandra module of an*

irreducible representation of G *on a Hilbert space.*

In (b), we may assume that K *acts by unitary operators.*

The proof of (a) is essentially a part of the proof of

Theorem 2.13(a). The known proofs of (b) all produce the

more explicit result known as Harish–Chandra's subquotient

theorem. This says that X must actually occur in a cer-

tain standard family of representations known as the princi-

pal series. The subquotient theorem is of enormous impor-

tance for establishing general properties of representations

(the moral equivalent of *a priori* estimates in the study of

differential equations), but has been of surprisingly little

value in the search for more detailed information (exact

solutions, to continue the analogy).

COROLLARY 2.16. *The set* $\hat{G}$ *of Definition 2.14 may be*

identified with the set of equivalence classes of irre-

ducible $(\mathfrak{g},K)$-*modules.*

To conclude this chapter, we give an algebraic

characterization of $\hat{G}_u$ as a subset of $\hat{G}$.

Definition 2.17. Suppose V is a $(\mathfrak{b},H)$-module. An *invariant Hermitian form* on V is a sesquilinear pairing $\langle\ ,\ \rangle$ from V to $\mathbb{C}$ satisfying

$$\langle h \cdot v, w \rangle = \langle v, (h^{-1}) \cdot w \rangle$$

$$\langle X \cdot v, w \rangle = -\langle v, X \cdot w \rangle,$$

for h in H, X in $\mathfrak{b}_0$, and v and w in V. The form $\langle\ ,\ \rangle$ is called *positive definite* if $\langle v, v \rangle$ is a positive real number for every vector v in V.

It is trivial to see that the inner product on any unitary representation of G induces a positive definite invariant Hermitian form on its Harish–Chandra module. The converse is less trivial, but still true:

PROPOSITION 2.18 (Harish–Chandra). *Suppose G is a reductive group, and K is a maximal compact subgroup. Assume that X is an irreducible $(\mathfrak{g},K)$-module admitting a positive definite invariant Hermitian form* (Definition 2.17). *Then X is the Harish-Chandra module of a unique irreducible unitary representation of G.*

Chapter 3

PARABOLIC INDUCTION

In this chapter, we will describe the first great success of unitary representation theory for reductive groups: the idea of parabolic induction.

Suppose to begin with that G is a locally compact group, and H is a closed subgroup. About 1950, Mackey showed how to use a unitary representation ϕ of a subgroup of H to construct a unitary representation

$$(3.1) \qquad \Phi = \text{Ind}_H^G(\phi)$$

of G. (Mackey's work is summarized in his book [Mackey, 1976]. In case the homogeneous space G/H carries a nice invariant measure, and ϕ is the trivial one-dimensional representation, then Φ is just the representation of G by left translation on $L^2(G/H)$. To motivate the general case, it is helpful to outline the proof of Proposition 1.20. Here is a restatement of it.

PROPOSITION 3.2. *Suppose* G *is a Lie group and* H *is a*

closed subgroup. Then there is a natural bijection between

(equivalence classes of) r-*dimensional homogeneous vector*

bundles on G/H, *and* r-*dimensional representations of the*

group H.

Proof. Suppose $\mathcal{V}$ is such a bundle, and π is the projec-

tion to G/H. To say that $\mathcal{V}$ is homogeneous means that

there is a continuous action of G on $\mathcal{V}$, compatible with

the action on G/H. In addition, we require that the action

of each element g restrict to a linear transformation from

the fiber $\mathcal{V}_x$ (which is $\pi^{-1}(x)$) to $\mathcal{V}_{g \cdot x}$. Write V for

the fiber $\mathcal{V}_{eH}$ over the identity coset of H. The preced-

ing condition shows that the action of any element of H

defines an endomorphism of V; it is easy to see that this

defines a representation of H on V.

Conversely, suppose that V carries a representation

of H. Define an equivalence relation $\sim$ on G×V by

(3.3) $(gh, v) \sim (g, \phi(h)v)$.

The quotient space of G×V by this equivalence relation is

written $G \times_H V$; it is easily seen to be a vector bundle on

G/H. (This uses the fact that G is locally a product of

H and a complementary submanifold. This is the only point

at which we need G to be a Lie group.) The left action of
G on G×V preserves ~, and so makes G ×$_H$ V a homogeneous
vector bundle. This construction inverts the preceding one.
Q.E.D.

The argument given here can be used to relate many possible properties of homogeneous vector bundles to properties
of representations. Propositions 1.19 and 1.21 are of this
nature. For another example, invariant Hermitian forms on
$\mathcal{V}$ (that is, G-invariant families of Hermitian forms on the
fibers) correspond to H-invariant Hermitian forms on V.

COROLLARY 3.4. *In the setting of the previous proposition,*
suppose (ϕ,V) *is a finite-dimensional representation of*
H, *and* $\mathcal{V}$ = G ×$_H$ V *is the corresponding vector bundle on*
G/H. *Then the space of* (smooth, continuous, *or* measurable)
sections of $\mathcal{V}$ *may be identified with the space of* (smooth,
continuous, *or* measurable) *functions* f *from* G *to* V,
satisfying

$$f(gh) = \phi(h^{-1})f(g)$$

for all g *in* G *and* h *in* H.

Proof. Let F be a section of $\mathcal{V}$. Define f by

$$f(g) = (g^{-1}) \cdot F(gH).$$

Since F(g) belongs to the fiber of $\mathscr{V}$ over gH, the right side of this formula belongs to the fiber over eH, which is V. The function so defined has the properties required by the proposition.

Conversely, suppose f is given. Define a map F_0 from G to G×V by

$$F_0(g) = (g, f(g)).$$

Then it is immediate from the definition in (3.3) that $F_0(gh) \sim F_0(g)$; so F_0 induces a map F from G/H to $\mathscr{V}$. It is easy to check that F is a section, and that this construction inverts the one above. Q.E.D.

In order to construct unitary representations, we need to be able to integrate sections of vector bundles in a translation-invariant way. The first difficulty with this is that G/H does not in general admit a translation-invariant measure. To deal with this, we need to recall a few facts about densities on manifolds.

Definition 3.5. Suppose V is a real vector space, of dimension m. A (real or complex) *density* on V is a (real or complex) multiple of Lebesgue measure on V. The space D(V) of densities on V is a one-dimensional vector space. Alternatively, one can think of a density as an equivalence class of pairs (ω,ϵ), with $\omega \in \Lambda^m(V^*)$ a (real or complex

valued) volume form on V, and ϵ an orientation of V.

The equivalence relation is

$$(\omega,\epsilon) \sim (-\omega,-\epsilon).$$

If T is an automorphism of V, we make T act on densi-

ties by the requirement

$$\int f(x) \ d(T\mu) = \int f(Tx) \ d\mu.$$

This makes T act on D(V) by the scalar $|\det T|^{-1}$.

Suppose M is an m-dimensional manifold. The *density*

bundle DM is the line bundle whose fiber at p is

$$D_pM = D(T_pM).$$

A *smooth density* on M is a section of the density bundle.

By the change of variable formula, a smooth density may be

identified with a (signed, or complex valued) measure on M,

which is a smooth multiple of Lebesgue measure on each coor-

dinate patch.

If d is a compactly supported section of DM, then

$$\int_M d$$

is well-defined; in the identification of densities with

measures, it is the total mass of M.

LEMMA 3.6.   *In the setting of Proposition 3.2, the (line)*

bundle of densities on G/H is induced by the character

$$\delta_{G/H}(h) = \left|\det(h \text{ acting on } (g/\mathfrak{h})^{*})\right|$$
$$= \left|\det(Ad_{\mathfrak{h}}(h))/\det(Ad_{g}(h))\right|^{-1}.$$

Sections of this bundle (at least compactly supported continuous ones) therefore have a well-defined integral, which is invariant under the action of G on sections.

This result is a consequence of the standard identification of the tangent space of G/H at eH with $g/\mathfrak{h}$. The function δ is called the *modular function* of G/H.

PROPOSITION 3.7. *Let* G *be a Lie group, and* H *a closed subgroup. Suppose* (ϕ, V) *is a finite-dimensional unitary representation of* H. *Recall from Lemma 3.6 the (positive real-valued) modular function* $\delta = \delta_{G/H}$. *Let* $\mathcal{V}$ *be the vector bundle on* G/H *induced by the representation* $\phi \otimes (\delta^{\frac{1}{2}})$ *of* H. *Then the space of compactly supported continuous sections of* $\mathcal{V}$ *admits a* G*-invariant pre-Hilbert space structure. (This structure is natural as soon as we fix an identification of the space of* $\delta^{\frac{1}{2}} \otimes \delta^{\frac{1}{2}}$ *with the densities on* $g/\mathfrak{h}$.

Proof. Suppose F_1 and F_2 are sections of $\mathcal{V}$. Let f_1 and f_2 be the corresponding V-valued functions on G

(Corollary 3.4). Define a complex-valued function ω on G
by

$$\omega(g) = \langle f_1(g), f_2(g) \rangle.$$

Because the operators $\phi(h)$ preserve this inner product, we
compute

$$
\begin{aligned}
\omega(gh) &= \langle f_1(gh), f_2(gh) \rangle \\
&= \langle \delta^{\frac{1}{2}}(h^{-1})\phi(h^{-1})f_1(g), \delta^{\frac{1}{2}}(h^{-1})\phi(h^{-1})f_2(g) \rangle \\
&= \delta(h^{-1})\langle f_1(g), f_2(g) \rangle.
\end{aligned}
$$

It follows that ω may be regarded as a section Ω of the
bundle on G/H induced by δ. By Lemma 3.6, this is the
density bundle. If F_1 and F_2 are compactly supported and
continuous, then so is Ω; so it can be integrated over
G/H. The integral is defined to be $\langle F_1, F_2 \rangle$. Q.E.D.

It is perhaps clear that the restriction to finite-
dimensional vector bundles has been made only to keep the
context as familiar as possible; our goal was only to motivate the following definition.

Definition 3.8 (Mackey). Suppose G is a locally compact
group, H is a closed subgroup, and (ϕ, V) is an irreducible unitary representation of H. There is a character $\delta =$
$\delta_{G/H}$ of H, with the following property: functions on G
satisfying

$$\omega(gh) = \delta(h^{-1})\omega(g)$$

have a translation-invariant integral "over G/H." (More

precisely, if ω is supported on a set of the form UH,

with U compact, and bounded and measurable on U, then

this integral is finite.) Speaking loosely, we write

$$\int_{G/H} \omega(x)dx$$

for this integral.

We now define a Hilbert space W, consisting of measur-

able functions f from G to V, satisfying the following two

conditions. First,

(3.9)(a) $f(gh) = \delta^{\frac{1}{2}}(h^{-1})\phi(h^{-1})f(g)$

for all g in G and h in H. Second,

(3.9)(b) $$\int_{G/H} \langle f(x),f(x)\rangle dx < \infty.$$

Arguing as in the proof of Proposition 3.7, we see that the

function

$$\omega(g) = \langle f(g),f(g)\rangle$$

has the right transformation property under H for the inte-

gral to be defined. Similarly, we see that an inner product

can be defined on W by

$$\langle f_1,f_2\rangle = \int_{G/H} \langle f_1(x),f_2(x)\rangle dx.$$

The group G acts on W by

$(3.9)(c)$ $(\Phi(g)f)(x) = f(g^{-1}x)$.

With this structure, W is a Hilbert space, and Φ is a unitary representation of G on W, the *induced representation* from H to G of ϕ:

$$\Phi = \text{Ind}_{H}^{G}(\phi).$$

Random induced representations are not often irreducible. As was hinted in the introduction, Mackey applied this definition to produce irreducible unitary representations of groups with a large normal subgroup N; the groups H from which he needed to induce typically contained N. In light of these facts, it is perhaps almost miraculous that large families of irreducible representations of reductive groups can be obtained by induction. This was discovered between the late 1940's and the middle 1950's, primarily by Gelfand and Naimark, Bruhat, and Harish-Chandra. Here are some necessary structural preliminaries.

Definition 3.10. Suppose $\mathfrak{g}$ is a complex reductive Lie algebra. Recall that a *Borel subalgebra* of $\mathfrak{g}$ is by definition a maximal solvable subalgebra of $\mathfrak{g}$; that all of these are conjugate under the group of inner automorphisms of $\mathfrak{g}$; and that one may be constructed from a Cartan subalgebra and a set of positive roots (Definition 1.14). A *parabolic subalgebra* $\mathfrak{p}$ of $\mathfrak{g}$ is one that contains a Borel subalgebra.

Such a subalgebra is necessarily equal to its own normalizer in $\mathfrak{g}$:

$$(3.11) \qquad \mathfrak{p} = \{X \in \mathfrak{g} \mid [X,\mathfrak{p}] \subset \mathfrak{p}\}.$$

The *nil radical* $\mathfrak{n}$ of $\mathfrak{p}$ is the largest nilpotent ideal in $\mathfrak{p}$. A *Levi factor* of $\mathfrak{p}$ is a reductive subalgebra $\mathfrak{l}$ of $\mathfrak{p}$ such that $\mathfrak{p} = \mathfrak{l}+\mathfrak{n}$; such a subalgebra always exists, and is unique up to conjugation by $\exp(\mathfrak{n})$.

Suppose G is a real reductive group, with Lie algebra $\mathfrak{g}_0$. A *parabolic subalgebra* $\mathfrak{p}_0$ of $\mathfrak{g}_0$ is one whose complexification is parabolic in $\mathfrak{g}$. The corresponding *parabolic subgroup* P of G is the normalizer of $\mathfrak{p}_0$ in G:

$$(3.12) \qquad P = \{g \in G \mid \mathrm{Ad}(g)(\mathfrak{p}) \subset \mathfrak{p}\}.$$

(See the comment below, however.) Because of (3.11), the Lie algebra of P is $\mathfrak{p}_0$, as the notation suggests. The *unipotent radical* of P is the (normal) subgroup $N = \exp(\mathfrak{n}_0)$. There exist Levi subalgebras of $\mathfrak{p}_0$; for example,

$$(3.13)(a) \qquad \mathfrak{l}_0 = \mathfrak{p}_0 \cap \theta\mathfrak{p}_0$$

is one. (Here θ is the Cartan involution defined in (2.2).) If $\mathfrak{l}_0$ is such a Levi subalgebra, the corresponding *Levi subgroup* L is the normalizer of $\mathfrak{l}_0$ in P. For the choice in (3.13)(a),

$$(3.13)(b) \qquad L = P \cap \theta P.$$

For disconnected groups, it is convenient to extend the
definition of parabolic subgroup slightly. Fix $\mathfrak{p}_0$, and
write P^+ for the group defined above. Set

$$(3.12)' \quad P^- = \{g \in P^+ \mid$$
$$Ad(g)\big|_{\mathfrak{p}} \text{ is an inner automorphism of } \mathfrak{p}\}.$$

Then a parabolic subgroup of G corresponding to P is
defined to be one between P^- and P^+.

Here is a basic source of parabolic subgroups. Suppose
$G(\mathbb{R})$ is the set of real points of a complex reductive algebraic group $G(\mathbb{C})$. Suppose $\mathscr{P}$ is a projective variety
defined over $\mathbb{R}$, such that $G(\mathbb{C})$ acts transitively on $\mathscr{P}(\mathbb{R})$,
and $G(\mathbb{R})$ acts transitively on $\mathscr{P}(\mathbb{R})$. Then the stabilizer in
$G(\mathbb{R})$ of a point of $\mathscr{P}(\mathbb{R})$ is a parabolic subgroup of $G(\mathbb{R})$.

Example 3.14. Suppose G is $GL(n)$, and V is the standard n-dimensional representation of G. Let $\mathscr{P}$ denote
the Grassman variety of k-dimensional subspaces of V.
This satisfies the hypotheses above. The stabilizer in G
of the standard k-dimensional subspace is the group

$$P = \left\{ \begin{bmatrix} A & B \\ 0 & D \end{bmatrix} \right\},$$

with A in GL(k), D in GL(n-k), and B any k by n-k matrix. The natural Levi factor L of P is the subgroup consisting of matrices with B equal to zero; it is isomorphic to GL(k) × GL(n-k).

To get all parabolic subgroups of GL(n), it suffices to replace the Grassmanian by any partial flag variety (consisting of increasing sequences of subspaces of V of specified dimensions).

For a slightly different example, let G be the group O(n,n) of linear transformations of $\mathbb{R}^{2n}$ preserving the quadratic form

$$Q(x_1,\ldots,x_{2n}) = (x_1)^2 + \ldots + (x_n)^2 - (x_{n+1})^2 - \ldots - (x_{2n})^2.$$

Let $\mathscr{G}$ be the variety of totally isotropic n-dimensional subspaces; that is, subspaces on which Q vanishes. Again the hypotheses are satisfied. We take as our base point the subspace

$$W = \{(v,v) \mid v \in \mathbb{R}^n\}.$$

Restriction of linear transformations to W provides an isomorphism of the Levi subgroup L of P with GL(n). We leave to the reader the task of describing P and L more explicitly.

Notice that in both of these examples, the standard maximal compact subgroup of $G(\mathbb{R})$ still acts transitively on $\mathcal{P}(\mathbb{R})$. (This requires a small argument in the second case.) This is a general phenomenon, described in the following lemma.

LEMMA 3.15. *Suppose* G *is a reductive Lie group,* K *is a maximal compact subgroup, and* P *is a parabolic subgroup of* G.

   a) G = KP. *That is, every element of* G *may be written (not uniquely) in the form* kp, *with* k *in* K *and* p *in* P.

   b) *The homogeneous space* G/P *may be identified with* $K/(K \cap P)$.

   c) *The modular function* δ *of* G/P *is trivial on* $K \cap P$; *so there is a* K-*invariant measure on* G/P. *If* ω *is a function on* G *transforming according to* δ *(cf. Definition 3.8), then*

$$\int_{G/P} \omega(x)dx = \int_{K/(K\cap P)} \omega(x)dx = \int_K \omega(K)dK.$$

Here is a little more useful structure.

Definition 3.16. Suppose P is a parabolic subgroup of the reductive Lie group G. Choose a Levi subgroup L of G as in (3.13). Recall from (2.1) the Cartan subspace s_0, the -1 eigenspace of θ. Define

$$a_0 = (\text{center of } \mathfrak{l}_0) \cap s_0$$
$$A = \exp(a_0)$$
(3.17)(a) $\quad s(m)_0 = \text{orthogonal complement of } a_0 \text{ in } s_0$
$$m_0 = (\mathfrak{l}_0 \cap \mathfrak{k}_0) \oplus s(m)_0$$
$$M = (L \cap K)\exp(s(m)_0)$$

The abelian group A is isomorphic by the exponential map to its Lie algebra; we call A a *vector group*. If we write N for the unipotent radical of P, then we have a *Langlands decomposition*

(3.17)(b) $\qquad\qquad P = MAN,$

a semidirect product with each factor normalizing the succeeding ones. In particular, L = MA. If G is connected, or is the group of real points of a connected algebraic group, then M commutes with A. In any case, M is a reductive group with compact center.

Because L is the semidirect product of M and A, and the identity component of M acts trivially on A, it

is very easy to describe the representations of L explicitly in terms of those of M and A. We leave this task to
the reader, however.

Definition 3.18. Suppose G is a reductive Lie group, and
$$P = LN = MAN$$
is a Langlands decomposition of a parabolic subgroup of G.
Let ϕ be a unitary representation of L. We can regard ϕ
as a representation of P, by making N act trivially; we may
occasionally denote this extended representation by $\phi \otimes 1$.
The *representation (parabolically) induced* from L to G
by ϕ is
$$\Phi = \text{Ind}_P^G(\phi \otimes 1).$$
The space of Φ will be written $\mathcal{H}_\Phi$. It consists of functions f from G to the space $\mathcal{H}_\phi$ of ϕ, satisfying the
following conditions:

   a) $f(gp) = \delta^{\frac{1}{2}}(p^{-1})\phi(p^{-1})f(g)$ (all p in P and g

in G); and

   b) $\int_{G/P} \langle f(x), f(x) \rangle dx < \infty.$

(Here $\delta = \delta_{G/P}$ is as in Lemma 3.6.) Because of Lemma
3.15, such functions are determined by their restrictions to
K. The restrictions must satisfy

   a)' $f(km) = \phi(m^{-1})f(k)$ (all m in K $\cap$ P and k in

K); and

b)' f belongs to $L^2(K, \mathcal{H}_\phi)$.

Conversely, any function K satisfying (a)' and (b)' extends uniquely to one on G satisfying (a) and (b).

Here are a few of the basic properties of parabolic induction.

THEOREM 3.19 [Bruhat, 1956], Harish-Chandra). *Suppose* G *is a reductive Lie group,* P = MAN *is a parabolic subgroup, and* ϕ *is an irreducible unitary representation of* MA. *Write*

$$\Phi = \text{Ind}(P \uparrow G)(\phi)$$

(Definition 3.18), a unitary representation of G.

  a) Φ *depends (up to equivalence) only on the* G-*conjugacy class of the pair* (MA,ϕ) *(rather than on* (P,ϕ)*).*

  b) Φ *is a direct sum of a finite number of irreducible representations.*

  c) For most ϕ, Φ *is irreducible.*

When P is a minimal parabolic subgroup, M is compact and ϕ is finite-dimensional. This is the setting in which the theorem was proved by Bruhat. His arguments are fairly direct, and are in the spirit of Mackey's analysis of irreducibility in the presence of normal subgroups. In the general case, some powerful machinery is needed – the ideas in Mackey's analysis seem to contribute almost nothing. (However, Gelfand and Naimark established some special cases of (c) by showing that Φ was already irreducible under some subgroup of G, to which the Mackey theory did apply. This idea is very powerful for GL(n), useful for the classical groups, and (to date) essentially useless for the exceptional groups.)

Parts (a) and (b) of the theorem follow easily from Harish-Chandra's theory of global characters (including the regularity theorem of [Harish-Chandra, 1965]). (The theory of intertwining operators of [Knapp-Stein, 1975] provides a more direct proof of (a).) Harish-Chandra obtained a precise form of (c) in unpublished work from the early 1970's. A very general sufficient condition for irreducibility (that is, another precise version of (c)) is implicit in Theorem 13.5, which comes from [Speh-Vogan, 1980].

The main point of Theorem 3.19 is that parabolic induction provides a way of building irreducible unitary representations of G from those of Levi subgroups. Of course one would like to know exactly how much of $\hat{G}_u$ can be constructed in this way from proper parabolic subgroups. A precise result in this direction will be given in Theorem 13.5.

Example 3.20. Let $P = MAN$ be a minimal parabolic subgroup of G. Then M and A commute, and M is compact. An irreducible unitary representation ϕ of MA is therefore of the form $\xi \otimes v$, with ξ an irreducible (finite-dimensional) unitary representation of M and v a unitary character of A. The induced representation Φ will be written $I(\xi \otimes v)$, or sometimes $I_P(\xi \otimes v)$; it is called a (*minimal*) *principal series representation*.

Let us analyze the parameter set for this series of representations of G a little more closely. The group M is in Harish–Chandra's class, because of $(3.12)'$. We can therefore parametrize its representations using Theorem 1.30. Fix a Cartan subgroup T of M (Definition 1.14), and define $H = TA$ (a direct product). Then

$$H = \{h \in G \mid Ad(h) \text{ is trivial on } \mathfrak{h}\}.$$

In analogy with Definition 1.28(b), we define $W = W(G,H)$ to be the normalizer of H in G, modulo H. This group acts on the set

$$\hat{H}_u = \hat{T} \times \hat{A}_u$$

of irreducible unitary representations of H. Now Theorem 1.30 allows us to associate to any character τ in $\hat{H}_u$ a principal series representation $I(\tau)$. By Theorem 3.19(a),

$$I(\tau) = I(w\tau).$$

We have not given a detailed discussion of the language
of direct integrals of representations of G, needed to discuss abstract harmonic analysis problems. Nevertheless, it
is worth knowing that we already have enough representations
to solve an interesting problem. We will therefore state
the result, dealing with our lack of machinery by being a
little vague.

THEOREM 3.21 ([Harish-Chandra, 1958]; cf. [Helgason, 1984]).
Suppose G *is a reductive Lie group in Harish-Chandra's*
class (Definition 0.6) and P = MAN *is a minimal parabolic*
subgroup. Define W *as in Example 3.20, and use the nota-*
tion there. Then

$$L^2(G/K) \cong \int_{\hat{A}_u/W} I(1 \otimes v)dv$$

Far more precise results about the action of G on functions on G/K are available. Even Harish-Chandra's original proof gave more than this, and the problem has been
studied intensively. (One can consult for example
[Helgason, 1984].) Yet all of the deeper results involve,
in more or less explicit ways, the representations I(1 ⊗ v);
and all are in some sense based on Theorem 3.21.

If G is complex (say for simplicity in Harish–Chandra's class), then the representations $I(\tau)$ of Example 3.20 suffice to decompose $L^2(G)$ as well. Even in that case, however, they are far from all the unitary representations of G. In the next chapter, we will explain how to push the idea of induction a little further, to get some rather different representations.

Chapter 4

STEIN COMPLEMENTARY SERIES AND THE

UNITARY DUAL OF GL(n,$\mathbb{C}$)

The notion of induced representation in Definition 3.8

depends on starting with a unitary representation; condition

3.9(b) in the definition of the space of the induced repre-

sentation makes no sense otherwise. However, the motiva-

tional material on vector bundles does not require a unitary

representation; so one should expect to be able to do some-

thing more generally. There are in general several possi-

bilities, depending on what kind of topological space is

wanted for the induced representation. In the case of para-

bolic induction, the following definition is convenient.

Definition 4.1. Suppose G is a reductive Lie group and

P = LN is a parabolic subgroup (Definition 3.10). Assume

that $(\phi, \mathcal{H}_\phi)$ is a representation of G on a Hilbert space such that the restriction of ϕ to K is unitary. Extend ϕ to all of P by making N act trivially. Define the *representation (parabolically) induced* from L to G,

$$\text{Ind}_P^G(\phi) = \Phi,$$

as follows. The space $\mathcal{H}_\Phi$ of Φ consists of functions f from G to $\mathcal{H}_\phi$, satisfying

a) $f(gp) = \delta^{\frac{1}{2}}(p^{-1})\phi(p^{-1})f(g)$ (all p in P and g in G); and

b)' the restriction of f to K lies in $L^2(K,\mathcal{H}_\phi)$.

Here $\delta = \delta_{G/P}$ is as in Lemma 3.6. We have called the second condition (b)' to emphasize the connection with the conditions in Definition 3.18. Just as in that special case, we may replace (a) by

a)' $f(km) = \phi(m^{-1})f(k)$ (all m in K ∩ P and k in K).

The space $\mathcal{H}_\Phi$ is therefore a Hilbert space, and K acts unitarily. In fact

(4.2) $[\text{Ind}_P^G(\phi)]_K \cong \text{Ind}_{K\cap P}^P(\phi|_{K\cap P})$.

The description (a)' of the representation space is called the *compact picture*.

We retain the notation of the definition. It will be useful to have explicit formulas for the representation in the compact picture. To that end, choose a minimal parabolic

subgroup

(4.3)(a) $P_{min} = M_{min}A_{min}N_{min}$

contained in P. (The Lie algebra α_0 of A will be a

maximal abelian subalgebra of the -1 eigenspace of θ and

will be contained in L.) Multiplication then defines a

diffeomorphism

(4.3)(b) $G \cong K \times A_{min} \times N_{min}$;

we write this as

(4.3)(b)' $G = KA_{min}N_{min}$,

the *Iwasawa decomposition* of G (see for example [Helgason,

1978]). We write κ, α and n for the coordinate func-

tions on G for this decomposition: any element g of G

is

(4.3)(c) $g = \kappa(g)a(g)n(g)$.

It is traditional to write H for the logarithm of a:

(4.3)(d) $H(g) \in \alpha_0$, $\exp(H(g)) = a(g)$.

Suppose now that f is a function satisfying (a)' and

(b)' of Definition 4.1. Then the extension of f to G

satisfying (a) of the definition is

(4.4) $f(g) = \delta^{1/2}(a(g)^{-1})\phi(a(g)^{-1})f(\kappa(g))$.

For x in K, one therefore calculates

(4.5) $[\Phi(g)f](x) =$

$$\phi(a(g^{-1}x)n(g^{-1}x))\{\delta^{\frac{1}{2}}(a(g^{-1}x))f(\ell(g^{-1}x))\}.$$

To understand (4.5), one should keep in mind the fact (contained in Lemmas 3.6 and 3.15) that, for fixed g, the map

$$x \rightarrow \kappa(g^{-1}x)$$

induces a diffeomorphism of $K/(K \cap P)$, with Jacobian

$\delta(a(g^{-1}x))$. Consequently, the part of (4.5) in braces is

just the unitary action on L^2 induced by this diffeomorphism.

We now have at our disposal a substantially larger set
of representations. For instance, Example 3.20 can be generalized to give a family of representations involving all
characters of A as parameters (not just unitary ones). Unfortunately, the new representations obtained in this way
are not given to us as unitary representations. Our goal is
to see that some of them are unitary anyway. To see that,
we now consider in abstract terms the ingredients needed to
make a representation unitary.

Definition 4.6. Suppose $(\pi,\mathcal{H})$ is a representation of G
on a Hilbert space. The Hermitian dual π^h of π is the
representation of G on the same Hilbert space $\mathcal{H}$ given by

$$\pi^h(g) = \pi(g^{-1})^*.$$

Here the star denotes adjoint of operators on $\mathcal{H}$, and is defined by

$$\langle Tv,w \rangle = \langle v,T^*w \rangle \qquad (v,w \text{ in } \mathcal{H}).$$

What we would like to do is replace the inner product
that is given to us, that is not preserved by π, by some
new inner product that is. Let us be cavalier about boundedness for a moment. Then any Hermitian form on $\mathcal{H}$ is of
the form

$$(4.7)(a) \qquad\qquad \langle v,w \rangle^A = \langle v,Aw \rangle,$$

for A some self-adjoint operator on $\mathcal{H}$. The condition
that the new inner product be G invariant may be written

$$(4.7)(b) \qquad\qquad A\pi(g) = \pi^h(g)A.$$

This is precisely the condition for A to *intertwine* π
and π^h; that is, to be a G-equivariant map from π to π^h.
Conversely, suppose A satisfies $(4.7)(b)$. It is easy to
check that A^* does as well. If π is irreducible, we
should therefore expect A^* to be a multiple of A. Replacing A by an appropriate multiple of itself, we get A to
be self-adjoint. Now $(4.7)(a)$ defines a G-invariant Hermitian form on $\mathcal{H}$. This discussion may be summarized as follows.

(False) PROPOSITION. *Suppose* π *is an irreducible represen-*
tation of G *on a Hilbert space. Then* π *admits a non-zero*
G-invariant Hermitian form if and only if π *is equivalent*

to π^h (Definition 4.6). *In that case, there is a self adjoint operator* A, *unique up to a real factor, that intertwines* π *and* π^h. *The representation* π *is equivalent to a unitary one if and only if the operator* A *is (positive or negative) definite.*

This proposition is false only because we were careless about boundedness. The machinery of Chapter 2 is designed to circumvent such problems. Using it, and the same formal reasoning, we arrive at

PROPOSITION 4.8. *Suppose* π *is an irreducible admissible representation of* G *on a Hilbert space and that the inner product is* K-*invariant. Then* π *is infinitesimally equivalent to a unitary representation if and only if there is a positive Hermitian operator*

$$A: \mathcal{H}_K \to \mathcal{H}_K,$$

that intertwines the $(\mathfrak{g},K)$-*module structures defined by* π *and* π^h.

Recall that we are trying to make unitary some non-unitarily induced repesentations. Evidently the next problem is to find operators A satisfying the requirements of

the proposition. First we identify the Hermitian duals π^h
in question.

PROPOSITION 4.9. *In the setting of Definition 4.1, the*
Hermitian dual of

$$\Phi = \text{Ind}_P^G(\phi)$$

is

$$\Phi^h = \text{Ind}_P^G(\phi^h).$$

That is, if the Hilbert space of $\text{Ind}(\phi)$ *is defined by*
(4.1)(a)' and (b)', then it coincides with the space of
$\text{Ind}(\phi^h)$; *and the operators of the representations satisfy*

$$\text{Ind}(\phi)(g) = [\text{Ind}(\phi^h)(g^{-1})]^*.$$

This is an easy consequence of the explicit formula (4.5)
for the operators and the remark following it. (Because of
the definition of the inner product in $\mathcal{H}_\Phi$, the proposition
is asserting the equality of two integrals over $K/(K \cap P)$.
What is used finally is the change of variables formula for
integrals).

The next problem is this: when is $\text{Ind}(\phi)$ infinitesimally equivalent to $\text{Ind}(\phi^h)$? For our purposes, enough equi-
alences of this nature will arise from Theorem 3.19(a). So
suppose $P = MAN$ is a parabolic subgroup of the reductive

group G. Assume for simplicity that G is in Harish-
Chandra's class; in particular, M and A commute. Fix an
irreducible unitary representation ξ of M. For v in
$\hat{A}$, set

(4.10)(a) $I_P(\xi \otimes v) = \text{Ind}_P^G(\xi \otimes v \otimes 1)$,

a series of representations parabolically induced from MA
to G. In accordance with Definition 4.1, we can regard all
of these representations as realized on the same Hilbert
space, with the same restriction to K. By Proposition 4.9,

(4.10)(b) $I_P(\xi \otimes v)^h = I_P(\xi \otimes (v^h))$.

By Lemma 1.2, we may identify $\hat{A}$ with $\mathfrak{a}^*$. In this identification, the Hermitian dual of a representation corresponds
the to negative complex conjugate of a linear functional; so

(4.10)(b)' $I_P(\xi \otimes v)^h = I_P(\xi \otimes (-\bar{v}))$.

At least in the case that $I_P(\xi \otimes v)$ is irreducible, Proposition 4.8 therefore says that we are seeking conditions under
which $I_P(\xi \otimes v)$ is equivalent to $I_P(\xi \otimes (-\bar{v}))$.

Fix an element $\tilde{w}$ in K, normalizing MA. Write w
for the image of $\tilde{w}$ in

$$W(G,MA) = N_K(MA)/(M \cap K).$$

Then w acts on $\hat{M}$ and (linearly) on $\mathfrak{a}^*$. Assume that

(4.10)(c) $w \cdot \xi \cong \xi$.

For *unitary* characters v of A, Theorem 3.19(a) now guarantees that

(4.10)(d) $I_p(\xi \otimes v) \cong I_p(\xi \otimes wv)$.

The next theorem meromorphically continues this equivalence to non-unitary v.

THEOREM 4.11 ([Knapp–Stein, 1980]). *In the setting just described, there is a rational family of intertwining operators*

$$\{A(w{:}v) \mid v \in \mathfrak{a}^*\},$$

with the following properties. Write $\mathcal{H}$ for the common Hilbert space of all the $I_p(\xi \otimes v)$.

   a) For v outside a countable locally finite union Z *of proper algebraic subvarieties of $\mathfrak{a}^*$, $A(w{:}v)$ is a linear isomorphism from $\mathcal{H}_K$ to itself.*

   b) For v not in Z, *$A(w{:}v)$ is an infinitesimal equivalence of $I_p(\xi \otimes v)$ with $I_p(\xi \otimes wv)$.*

   c) Any K-finite matrix entry of $A(w{:}v)$ is a rational function of v.

   d) If $w^2 \cdot v = v$, and v is not in Z, *then*
$$A(w{:}v)^* = A(w{:}{-}\overline{w}\overline{v}).$$
In particular, if $wv = -\overline{v}$, then $A(w{:}v)$ is Hermitian.

   e) If $I_p(\xi \otimes v)$ is irreducible, then v is not in Z.

Knapp and Stein prove this result by constructing the intertwining operators as explicit integral operators for certain
v, then proving that they can be continued meromorphically.
This approach leads to a wealth of detailed information
about the operators and their connections with harmonic analysis. We will be content with the properties listed here,
however. For that, a fairly easy non-constructive proof can
be given. Here is an outline of it.

Sketch of proof. We begin by writing explicitly the intertwining conditions being imposed. They are

(4.12)(a) $A(w{:}v)I_p(\xi{\otimes}v)(k) = I_p(\xi{\otimes}wv)(k)A(w{:}v)$ $(k \in K)$

(4.12)(b) $A(w{:}v)I_p(\xi{\otimes}v)(Z) = I_p(\xi{\otimes}wv)(Z)A(w{:}v)$ $(Z \in \mathfrak{g})$.

Of course (4.12)(b) may be replaced by

(4.12)(b)' $A(w{:}v)I_p(\xi{\otimes}v)(u) = I_p(\xi{\otimes}wv)(u)A(w{:}v)$ $(u \in U(\mathfrak{g}))$.

For simplicity, assume that there is an irreducible representation μ of K that occurs with multiplicity one in
$\mathcal{H}$. Write $\mathcal{I}$ for the set of v such that $I_p(\xi \otimes v)$ is irreducible, and $\mathcal{I}_0$ for the intersection of $\mathcal{I}$ with $(i\alpha_0)^*$.
It is known that $\mathcal{I}$ is the complement of a locally finite
union R of algebraic subvarieties in α^*. (Because of
(e), R will turn out to contain Z; it will almost always
be strictly larger.) In particular, $\mathcal{I}_0$ is Zariski dense

in α^*. For v in $\mathcal{F}_0$, define $A(w:v)$ to be the unique
infinitesimal equivalence from $I_P(\xi \otimes v)$ to $I_P(\xi \otimes wv)$
that restricts to the identity on $\mathcal{H}_\mu$. The main point is to
prove (c) for v in $\mathcal{F}_0$; this makes sense since $\mathcal{F}_0$ is
Zariski dense. This fact in turn depends on

LEMMA 4.13. *Fix* Z *in* $\mathfrak{g}$. *In the setting* (4.10), *the action of* Z *in* $\mathcal{H}$ *depends on the parameter* v *in an affine
way. More precisely, choose a basis* $X_1, \dots, X_r$ *of* α.
Then there are operators $T_0, T_1, \dots, T_r$ *on* $\mathcal{H}_K$ *(depending
linearly on* Z), *such that*

$$I_P(\xi \otimes v)(Z) = T_0 + \sum_{i=1}^{r} v(X_i)T_i$$

The lemma may be proved in a straightforward way by differen-
tiating (4.5); we omit the details.

COROLLARY 4.14. *In the setting of Theorem* 4.11, *fix a basis*
$\{v_i\}$ *of* $\mathcal{H}_\mu$. *Suppose* $\{w_j\}$ *is any other finite set of vectors in* H_K. *Then we can find elements* $\{u_{ji}\}$ *of* $U(\mathfrak{g})$,
depending rationally on v, *so that*

$$w_j = \sum_i u_{ji}v_i$$

for all v *for which the rational functions are defined.*

Sketch of proof. Possibly after expanding the set of w_j somewhat, one can find finitely many elements q_k of $U(\mathfrak{g})$, such that

$$(4.15) \qquad q_k v_i = \sum f_{ikj}(\upsilon) w_j,$$

with the coefficients f polynomial in υ. Because of the irreducibility of $I_P(\xi \otimes \upsilon)$, we can arrange for the original w's to be in the span of the various $q_k v_i$ (for most υ). Solving (4.15) by row reduction gives the corollary. $\square$

The fact that $A(w{:}\upsilon)$ depends rationally on υ is now a formal consequence of the intertwining condition $(4.12)(b)'$, Corollary 4.14, and Lemma 4.13. This is Theorem 4.11(c).

Because $A(w{:}\upsilon)$ commutes with K, it preserves the decomposition of $\mathcal{H}_K$ into K-primary subspaces (cf. (2.8)); we write

$$(4.16) \qquad A(w{:}\upsilon) = \sum A(w{:}\upsilon)_\delta$$

accordingly. Each of the summands is now a rational function from α^* to $\text{End}(\mathcal{H}_\delta)$. Define Z to be the union over δ of the poles of $A(w{:}\upsilon)_\delta$, and the zeros of $\det(A(w{:}\upsilon)_\delta)$. Obviously Z satisfies the last assertion of Theorem 4.11(a); that the union is locally finite will follow from (e) and the remarks after (4.12).

By Theorem 4.11(c) and Lemma 4.13, both sides of the

equations (4.12) depend rationally on v. Since they are

true by definition on the Zariski dense set $\mathcal{P}_0$, they are

true everywhere. This is Theorem 4.11(b).

A formal argument shows that the adjoint $A(w{:}v)^*$

intertwines $I_P(\xi \otimes (-wv))$ and $I_P(\xi \otimes (-\bar{v}))$. Furthermore,

the adjoint is still the identity on $\mathcal{H}_\mu$. By definition, it

therefore agrees with $A(w{:}(-w\bar{v}))$ for v in $\mathcal{P}_0$ such that

$w(-w\bar{v})$ is equal to $-\bar{v}$. This latter condition is equiva-

lent to $w^2{\cdot}v = v$, proving Theorem 4.11(d) on $\mathcal{P}^0$. The gen-

eral case follows by a density argument.

The proof of Theorem 4.11(e) requires a little care.

Suppose v_0 is not in Z; we want to prove that $I_P(\xi \otimes v_0)$

is reducible. If $A(w{:}v_0)$ is well defined, then (by the

definition of Z), it must not be invertible. Its kernel is

therefore a proper invariant subspace of $I_P(\xi \otimes v_0)$. (The

kernel cannot be all of $\mathcal{H}_K$, because it does not meet $\mathcal{H}_\mu$.)

We may therefore assume that $A(w{:}v)$ has a pole at v_0. It

is natural to consider the subspace on which $A(w{:}v)$ is

finite at v_0, and to think that it should provide a proper

invariant subspace of $\mathcal{H}_K$. It is not invariant, however, as

a careful attempt to write the obvious argument reveals.

The correct approach is to define

(4.17) $(\mathcal{H}_K)_0 = \{v \in \mathcal{H}_K \mid$ there is a polynomial function f

of v such that $f(v_0) = v$, and $A(w\!:\!v)f(v)$ is

finite at $v_0\}$

This is a proper $I_p(\xi \otimes v_0)$-invariant subspace, proving

(e).   □

The definition in (4.17) can be generalized, giving a

family of invariant subspaces parametrized by the submodules

of the local ring at v_0 in its quotient field. This fami-

ly is called the *Jantzen filtration*. It plays an important

part in representation theory (cf. [Jantzen, 1979] and

[Vogan, 1984], for example) but is very far from being well

understood.

Our efforts in the proof of Theorem 4.11 are now reward-

ed by unitary representations growing on trees (or at any

rate on crosses).

THEOREM 4.18. *In the setting (4.10), assume that* $I_p(\xi \otimes 0)$

is irreducible. Define a real subspace S_w *of* $\mathfrak{a}^*$ *by*

$$S_w = \{v \mid wv = -\bar{v}\}.$$

Write R *for the set of* v *for which* $I_p(\xi \otimes v)$ *is reduci-*

ble. Finally, let C_w *denote the connected component of*

the origin in $S_w - R$. *Then* $I_p(\xi \otimes v)$ *is infinitesimally equivalent to an irreducible unitary representation of* G

for every v *in* C_w.

Proof. Fix v in C_w. By Theorem 4.11, $A(w:v)$ is a Hermitian intertwining operator from $I_p(\xi \otimes v)$ to $I_p(\xi \otimes wv)$;
it depends rationally on v. By the "only if" part of Proposition 4.8, $A(w:0)$ must be definite. By continuity,
$A(w:v)$ must be definite as well. The theorem now follows
from the "if" part of Proposition 4.8.   □

The representations constructed in Theorem 4.18 are
called *complementary series*. The term is also applied in a
variety of generalizations; for this style of argument can
be pushed much further. Unitarity occurs either everywhere
or nowhere on a connected component of $S_w - R$. It often
happens that such a component will contain a few points
where unitarity is easier to prove or disprove. One can get
an immediate improvement by replacing R with the set Z of
Theorem 4.11. I have used R only because it is sometimes
easier to compute. Another possibility is to investigate
the nature of the poles and zeros of $A(w:v)$ along R, and
to use this information to study the positivity of A.
(Zeros of even order do not affect positivity, for example.)

For some hints about how complicated this study can become, the reader may consult [Knapp–Speh, 1983] or [Duflo, 1979].

Here is a basic example. Suppose G is the general linear group $GL(2m,\mathbb{C})$ of invertible $2m{\times}2m$ matrices. We can take as a Cartan involution the map

(4.19)(a) $\theta g = (g^{-1})^{*}$,

the inverse conjugate transpose. Then

(4.19)(b) $K = U(2m)$,

the group of unitary operators on $\mathbb{C}^{2m}$. Let P be the parabolic subgroup of Example 3.14, with $k = m$; as discussed there, the Levi factor MA is $GL(m,\mathbb{C}){\times}GL(m,\mathbb{C})$. To discuss such groups, it is convenient to use the following notation: if $n = p_1+\ldots+p_r$, and B_i is a $p_i{\times}p_i$ square matrix, then

$$d(B_1,\ldots,B_r)$$

denotes the block–diagonal matrix with the indicated diagonal blocks, and zeros elsewhere. We will also write I_p for the $p{\times}p$ identity matrix. Then the group A is a product of two copies of $\mathbb{R}$, by

(4.19)(c) $A = d(e^{t_1}I_m, e^{t_2}I_m)$ $(t \in \mathbb{R})$.

Similarly, M is a product of two copies of what might be (but never is) called $UL(m,\mathbb{C})$:

(4.19)(d) $M = d(m_1, m_2)$ (m_i an m×m matrix, $|\det(m_i)| = 1$).

The group M has unitary characters $\xi(k_1, k_2)$ (k_i in $\mathbb{Z}$),
defined by

(4.20)(a) $\xi(k_1, k_2)(d(m_1, m_2) = [\det(m_1)]^{k_1}[\det(m_2)]^{k_2}$.

Similarly, the characters of A are parametrized by $\mathbb{C}^2$, by

(4.20)(b) $(\upsilon_1, \upsilon_2)(d(x_1, x_2)) = [\det(x_1)]^{\upsilon_1}[\det(x_2)]^{\mu_2}$.

In these coordinates, the modular function δ for G/P is
trivial on M, and on A is given by

(4.20)(c) $\delta = (2m, -2m)$.

    Consider now the element

$$\tilde{w} = \begin{bmatrix} 0 & I_m \\ -I_m & 0 \end{bmatrix}$$

of $GL(2m, \mathbb{C})$. (We use the minus sign only to put $\tilde{w}$ in
$SL(2m, \mathbb{C})$. For our purposes that is unnecessary, but the
possibility of doing so shows that the construction works
for SL as well as for GL.) This element belongs to K
and normalizes MA; it acts there by permuting the two
$GL(m, \mathbb{C})$ factors. Write w for the image of $\tilde{w}$ in $W(G, MA)$
(cf.(4.10)). Then

(4.21)(a) $w(\xi(k_1, k_2)) = \xi(k_2, k_1)$.

To be in the setting of (4.10), we should therefore restrict
attention to the case $k_1 = k_2$. Similarly, one sees that
the set S_w of Theorem 4.18 is

(4.21)(b) $S_w = \{(\sigma+it,-\sigma+it)\}.$

Finally, we need to know the set R of reducibility points for the non-unitarily induced representations. We will take the result as a gift from non-unitary representation theory, without discussing a proof.

LEMMA 4.22. *In the setting (4.19)-(4.21), the induced representation*

$$I_P(\xi(k,k)\otimes v)$$

is reducible if and only if v_1-v_2 *is a non-zero even integer.*

Now we can apply Theorem 4.18, to get

THEOREM 4.23 ([Stein, 1967]). *Suppose* $G = GL(2m,\mathbb{C})$. *Fix notation as in (4.19)-(4.22). For* k *in* $\mathbb{Z}$, t *in* $\mathbb{R}$, *and* σ *strictly between* -1 *and* 1, *the representation*

$$C_{2m}(k,t:\sigma) = Ind_P^G(\xi(k,k) \otimes (\sigma+it,-\sigma+it) \otimes 1)$$

is (infinitesimally equivalent to) an irreducible unitary representation of G.

For σ equal to zero, these representations are unitarily induced. For σ not equal to zero, they are called the

Stein complementary series for $GL(2m,\mathbb{C})$. The same construction applies over any local field $\mathbb{F}$. The parameter space
will be the product of an open interval with the set of unitary characters of $\mathbb{F}^{\times}$.

A few remarks about the parameters may clarify at least
the notation slightly. The unitary characters of $GL(n,\mathbb{C})$
are parametrized by $\mathbb{Z}{\times}\mathbb{R}$, by the rule

$(4.24)(a)$ $\chi(k,t)(g) = [\det(g)/|\det(g)|]^{k}[|\det(g)|]^{it}$.

Inspection of the definitions shows that

$(4.24)(b)$ $C_{2m}(k,t{:}\sigma) \cong C_{2m}(0,0{:}\sigma) \otimes \chi(k,t)$.

The dependence on k and t is therefore uninteresting.
In addition, the construction of the inner product required
an isomorphism

$(4.24)(c)$ $C_{2m}(k,t{:}\sigma) \cong C_{2m}(k,t{:}-\sigma)$.

We may therefore confine attention to σ in the interval
$(0,1)$.

In view of the way Theorem 4.18 is proved, it may also
be helpful to understand a little about the behavior of the
induced representations as the continuous parameter varies.
For z in $\mathbb{C}$, put

(4.25) $I(z) = \operatorname{Ind}_{P}^{G}(\xi(0,0) \otimes (z,-z) \otimes 1)$.

Lemma 4.22 says that $I(z)$ is irreducible unless z is a
non-zero integer. We therefore concentrate on those repre-

sentations. By the definition of induction and (4.20)(c),

I(-m) is the space of functions on G/P. It therefore con-

tains a copy of the trivial representation of G, on the

constant functions. Dually, I(m) is the space of densi-

ties on G/P. Elements of it may be integrated; those with

integral zero form an invariant subspace. The quotient of

I(m) by this subspace is the trivial representation of G.

For k an integer between 0 and m (inclusive), let

P_k be the parabolic subgroup of Example 3.14. Set

$$(4.26) \qquad\qquad J(k) = \text{Ind}_{P_{m-k}}^{G} (1).$$

Then it turns out that J(k) is a subrepresentation of

I(-k) and a quotient of I(k). This is trivial for k = 0,

and we have just checked it for k = m. The intermediate

cases are not so easy. Although all but the last are

infinite-dimensional, the representations J(k) decrease in

size as k increases, in a certain precise sense.

For any non-negative integer r, let

$$(4.27) \qquad K(r) = \text{Ind}_{P}^{G}(\xi(r,r) \otimes (0,0) \otimes 1).$$

Then K(r) is a subrepresentation of I(r) and a quotient

of I(-r). Again this is trivial for r = 0, but requires

some proof for r positive. For r = 0, J(0) = K(0). For

r = 1, J(1) and K(1) are the only two composition fac-

tors of I(±1). For r greater than one, there are gener-

ally additional composition factors.

The trivial representation of K occurs in $I(z)$ with multiplicity one. Let us normalize the intertwining operator $A(w:z)$ (which takes $I(z)$ to $I(-z)$) to be the identity on the trivial representation of k. Then $A(w:z)$ has poles exactly at the negative integers, and zeros exactly at the positive integers. (By a zero, we mean that the operator has a kernel; it is still non-trivial). For k between 0 and m, the image of $A(w:k)$ is the subrepresentation $J(k)$ of $I(-k)$, described above. For r positive, the quotient of $I(-r)$ by the invariant subspace on which $A(w:-r)$ has a pole of less than maximal order (cf. (4.17) and the remarks following it) is $K(r)$. (Another way to say this is that if we kill the pole by renormalizing, the image of $A(w:-r)$ is $K(r)$.)

There is much more to be said about $I(z)$, but these are some of the highlights. Further results, particularly analytic ones, can be found in [Stein, 1967].

Miraculously, even this most primitive part of the theory of complementary series provides all the unitary representations of $GL(n,\mathbb{C})$.

THEOREM 4.28 ([Vogan, 1986b]). *Suppose* G *is* $GL(n,\mathbb{C})$. *Let* π *be an irreducible unitary representation of* G. *Then we can find a parabolic subgroup* $P = LN$ *of* G, *and an irreducible unitary representation* ϕ *of* L, *with the following properties. Write* L *as a product of various* $GL(p_i, \mathbb{C})$ *(as is always possible).*

a) ϕ *is a tensor product of irredcible unitary representations* ϕ_i *of* $GL(p_i, \mathbb{C})$.

b) $\mathrm{Ind}_P^G(\phi) \cong \pi$.

c) *Either* ϕ_i *is a (one-dimensional) unitary character of* $GL(p_i, \mathbb{C})$ *(cf.* (4.24)(a)); *or* p_i *is even, and* ϕ_i *is a Stein complementary series representation (Theorem* 4.23).

Conversely, if P = LN *is a parabolic subgroup of* G *and* ϕ *is any irreducible unitary representation of* L *satisfying* (c), *then* $\mathrm{Ind}_P^G(\phi)$ *is an irreducible unitary representation of* G.

Finally, the only equivalences among these induced representations are those provided by Theorem 3.19(a). *That is,* π *determines the conjugacy class of* (L, ϕ) *under* G.

The last two assertions (about irreducibility and inequivalence) were essentially proved by Gelfand and Naimark more than thirty years ago; they are not very difficult. Some hints about how to prove the main assertions may be found in Chapter 13.

We conclude this chapter with a simple corollary of Theorem 4.28.

COROLLARY 4.29. *Suppose* G *is* $GL(n, \mathbb{C})$, $P = LN$ *is a parabolic subgroup, and* ϕ *is an irreducible unitary representation of* L. *Then* $\mathrm{Ind}_P^G(\phi)$ *is irreducible.*

Chapter 5

COHOMOLOGICAL PARABOLIC INDUCTION: ANALYTIC THEORY

One of Harish-Chandra's fundamental insights was that representations of real reductive groups should be parametrized approximately by characters of Cartan subgroups. As we saw in Chapter 1, the Cartan-Weyl theory accomplishes this when G is compact. For general G, parabolic induction provides representations associated to characters of one Cartan subgroup (cf. Example 3.20). If G is complex, there is only one conjugacy class of Cartan subgroups; so Harish-Chandra's idea is completely implemented. Most real groups have several conjugacy classes of Cartan subgroups, however; so parabolic induction is not enough. Harish-Chandra's successful efforts (culminating in [Harish-Chandra, 1966]) to overcome this problem lie at the heart of his work, and constitute one of the great achievements of modern mathematics. Using almost nothing but the ordinary abelian Fourier transform, he constructed by hand certain

spaces of functions on G; the necessary representations

appeared as the action by translation on these spaces of

functions. The simplest case is a construction of the spher-

ical harmonics on S^3 (which may be thought of as the group

SU(2).) Roughly speaking, he constructs a function first on

the tangent space at a point, as the Fourier transform of

the invariant measure on a (two-dimensional) sphere in the

cotangent space. Under appropriate conditions, this can be

lifted to S^3 by the exponential map. This gives a zonal

spherical function; others can be obtained by letting SO(4)

act. Already in this case, the construction is ingenious,

and not trivial to carry out. For general G, the difficul-

ties are multiplied; I will not try even to outline the

ideas, or the deep theorems needed to implement them. Twenty

years of effort have not yet completely integrated this mate-

rial with the rest of mathematics (in the usual sense of

providing generalizations, proofs by standard techniques,

and so on).

   Fortunately for mortals (or at any rate for mortal alge-

braists), these twenty years *have* produced several alterna-

tive approaches to the basic problem of finding some more

representations. This problem can be regarded as that of

solving some elliptic linear differential equations. Harish-

Chandra essentially writes formulas for the solutions; but
if we are willing to forego these, we can hope to have less
trouble. In [Atiyah–Schmid, 1977], the index theorem is
used to guarantee the existence of some solutions. (Because
the domain is essentially G itself, which is non–compact,
this is by no means an easy exercise. Roughly speaking, the
substitute for compactness is in the G–invariance of the
problem.) A second approach is that of Flensted–Jensen.
There the idea is to obtain the required functions by a very
clever analytic continuation from those related to principal
series representations on $G_{\mathbb{C}}$. Although this is perhaps the
least obvious of the known methods, it may be technically
the simplest to carry out. A detailed account of it is in
[Knapp, 1986].

We will adopt a third approach. Before discussing it
in detail, we need some terminology.

Definition 5.1. Suppose G is a real reductive Lie group.
A subalgebra $\mathfrak{l}_0$ of $\mathfrak{g}_0$ is called a *Levi subalgebra* if its
complexification $\mathfrak{l}$ is a Levi factor of a parabolic subalge-
bra $\mathfrak{q} = \mathfrak{l} + \mathfrak{u}$ of $\mathfrak{g}$ (see Definition 3.10). (We do *not*
require that $\mathfrak{q}$ should be defined over $\mathbb{R}$.) Equivalently,
a Levi subalgebra is one that is the centralizer in $\mathfrak{g}_0$ of
some semisimple element of $\mathfrak{g}_0$.

Fix a Levi subalgebra $\mathfrak{l}_0$ and a corresponding parabolic $\mathfrak{q}$. Set

(5.2)(a) $L^+ = \{x \in G \mid Ad(x)$ preserves $\mathfrak{l}_0$ and $\mathfrak{q}\}$,

the *large Levi subgroup* attached to $\mathfrak{l}_0$ and $\mathfrak{q}$. Next,
define the *small Levi subgroup* attached to $\mathfrak{l}_0$, by

$$(5.2)(b)\quad L^- = \{x \in G \mid Ad(x) \text{ is inner on } \mathfrak{l}_0 \text{ and } \mathfrak{q}\}$$
$$= \{x \in L^+ \mid Ad(x) \text{ is inner on } \mathfrak{l}\}$$

$$= \text{largest subgroup of } L^+ \text{ in}$$

$$\text{Harish--Chandra's class}$$

$$= L^+ \cap G^-.$$

Here G^- is defined by any of the first three equations,
applied to the case $\mathfrak{l} = \mathfrak{q} = \mathfrak{g}$. The equivalence of these
four definitions is a fairly easy exercise in structure
theory for reductive Lie algebras. Finally, a *general Levi
subgroup* attached to $\mathfrak{q}$ and $\mathfrak{l}_0$ is one between L^- and
L^+. (This notation is consistent with that given for com-
pact groups in Definition 1.28 and for Levi factors of real
parabolics in Definition 3.10.)

Suppose L is a Levi subgroup. The *Weyl group* of L
in G is

$$W(G,L) = N_G(L)/L,$$

the quotient of the normalizer of G in L by L. We will
be interested in this almost exclusively in the case when L
is small. Similarly, define

$$W(\mathfrak{g},\mathfrak{l}) = W(G_{\mathbb{C}},L_{\mathbb{C}}).$$

Here $G_{\mathbb{C}}$ is any complex connected group with Lie algebra $\mathfrak{g}$, and $L_{\mathbb{C}}$ is the connected subgroup corresponding to $\mathfrak{l}$. If G is in Harish-chandra's class, then $W(\mathfrak{g},\mathfrak{l})$ contains $W(G,L)$; but this fails in general.

A *Cartan subgroup* of G is a Levi subgroup H of minimal dimension (called the *rank* of G). For Levi subgroups, the minimal dimension assumption is equivalent to requiring that $\mathfrak{h}$ be abelian, or that the image of $Ad(H)$ in $End(\mathfrak{h})$ be finite. In this case, the Weyl group $W(\mathfrak{g},\mathfrak{h})$ is the Weyl group of the root system of $\mathfrak{h}$ in $\mathfrak{g}$ (cf. [Humphreys, 1972]).

In the notation of the definition, G is in Harish-Chandra's class if and only if $G^- = G$. In that case the distinction between small and large Levi factors disappears.

The main general construction of Levi subgroups is this: if $\mathfrak{z}_0$ is an abelian subalgebra of $\mathfrak{g}_0$ consisting of semisimple elements, then

(5.3) $L = \{x \in G \mid Ad(x)\big|_{\mathfrak{z}} \text{ trivial}\}$

is a Levi subgroup. As an example, suppose G is $GL(2n,\mathbb{R})$. The Lie algebra of G may be identified with the space of endomorphisms of $\mathbb{R}^{2n}$. Identify $\mathbb{R}^{2n}$ with $\mathbb{C}^n$, and let X be the endomorphism of multiplication by i. The central-

izer L of X in G consists of real linear transformations of $\mathbb{C}^n$ that commute with multiplication by i; that
is, of complex linear transformations. Consequently

(5.4) $L = GL(n,\mathbb{C}) \subset GL(2n,\mathbb{R})$

is a Levi factor. Using similar ideas, the reader may
easily exhibit $U(r){\times}SO(n-r,n-r)$ as a Levi factor in
$SO(n,n)$. To understand the technical problem associated
with small and large Levi factors, keep in mind that
$SO(2){\times}SO(2)$ is a small Cartan subgroup of $O(4)$, and
$SO(2)xO(2)$ is a large one.

Here is a helpful structural fact.

LEMMA 5.5. *Any Levi subgroup of a reductive group* G *is*
conjugate to a θ-*stable one. Two* θ-*stable Levi subgroups*
are conjugate by G *if and only if they are conjugate by* K.

Any θ-stable Levi factor L has a Langlands decomposition

(5.6) $L = MA$,

with the properties discussed in Definition 3.16. In particular, $\mathfrak{a}_0$ is contained in the -1 eigenspace of θ, and the
centralizer of $\mathfrak{m}_0$ in M is compact. If L is small,
then A is central in L.

We will also need an appropriate extension of Definition 1.33.

Definition 5.7. Suppose L is a Levi subgroup of G and $q = I + u$ is a parabolic subalgebra normalized by L. Write $2\rho(u)$ for the determinant character of L on u:

$$2\rho(u)(x) = \det(\mathrm{Ad}(x)\big|_u).$$

The differential of $2\rho(u)$ is denoted by the same symbol. Its restriction to any Cartan subalgebra $\mathfrak{h}$ of I is the sum of the roots of $\mathfrak{h}$ in u. The *metaplectic cover* $L^{\sim}$ of L is the two-fold cover defined by the square root $\rho(u)$ of $2\rho(u)$. We will generally write ζ for the nontrivial element of the kernel of the covering map. A *metaplectic representation* of $L^{\sim}$ is one that is -1 on ζ.

Following the line of reasoning given after Proposition 1.35, one can check that $L^{\sim}$ depends only on L, and not on the particular q used to define it.

LEMMA 5.8. *Suppose* $\mathfrak{s}_0$ *is an abelian subspace of the* -1 *eigenspace of* θ *on* $\mathfrak{g}_0$. *Then the centralizer* L *of* $\mathfrak{s}_0$ *in* G *is the Levi factor of a real parabolic subgroup of* G.

Sketch of proof. Let X be a generic element of $\mathfrak{s}_0$. The eigenvalues of ad(X) on $\mathfrak{g}$ are all real; so it makes sense to define

$\mathfrak{p}_0$ = sum of the non-negative eigenspaces of ad(X).

This will be a parabolic subalgebra, with Levi subalgebraequal to the centralizer of X in $\mathfrak{g}_0$. Since X was chosen to be generic, this centralizer is just the Lie alge-bra $\mathfrak{l}_0$ of L. The rest of the argument is easy, and we omit it.   □

We turn now to a continuation of the discussion at the beginning of this chapter, on associating representations to characters of Cartan subgroups. So suppose that H = TA is a Langlands decomposition of a θ-stable Cartan subgroup of G. Recall that this means that T is contained in K, and A is a vector group. By Lemma 5.8, the centralizer of A in G is the Levi factor of some parabolic subgroup P = MAN of G. Necessarily M contains T_0, and T ∩ M is a compact Cartan subgroup of M. Assume that T and A com-mute. (This is automatic if G is in Harish-Chandra's class, but it excludes some interesting behavior in gener-

al.) Then T itself is a compact Cartan in M.

We are looking for a series of representations of G

parametrized by representations of H. Harish-Chandra pro-

posed to construct them in analogy with Example 3.20, as

follows. Any irreducible unitary representation τ of H

is of the form $\mu \otimes \upsilon$, with μ an irreducible unitary repre-

sentation of T, and υ a unitary character of A. Suppose

we have some way to associate to μ a unitary representa-

tion $I(M{:}\mu)$ of M. Then we can define

$$(5.9) \qquad I(G{:}\tau) = \mathrm{Ind}_P^G(I(M{:}\mu) \otimes \upsilon \otimes 1),$$

a unitary representation of G.

In this way, one is led to concentrate on the case of a

compact Cartan subgroup. To state what Harish-Chandra did

about this case, we need one more definition.

Definition 5.10. Suppose G is a locally compact unimodular

group. An irreducible unitary representation π of G is

said to belong to the *discrete series* if it is equivalent to

a subrepresentation of the translation representation of G

on $L^2(G)$. An equivalent condition is that for any v and

w in the space of π, the matrix coefficient $\langle \pi(g)v,w \rangle$

should be square integrable as a function on G.

If G is compact, every irreducible unitary representation is in the discrete series. Harish-Chandra's result is
an extension to non-compact G of Theorem 1.37.

THEOREM 5.11 ([Harish-Chandra, 1966]). *Suppose* G *is a*
reductive Lie group. The representations of the discrete
series of G *are parametrized by* G*-conjugacy classes of*
pairs (T,τ), *where*

a) T *is a large compact Cartan subgroup of* G, *normal-*
izing a Borel subalgebra $\mathfrak{b}$ *of* $\mathfrak{g}$; *and*

b) τ *is a dominant regular metaplectic representation*
of $\tilde{T}$ *(Definition 5.7 and Definition 1.36).*
In particular, the discrete series is non-empty if and only
if G *has a compact Cartan subgroup.*

We write I(τ), or $I_{\mathfrak{b},T}(G{:}\tau)$, for the discrete series with
parameter τ.

The result may also be formulated a little more along
the lines of Theorem 1.30, as follows.

THEOREM 5.12 ([Harish-Chandra, 1966]). *Suppose* G *is a*
reductive Lie group and T *is a small Cartan subgroup of*
G. *Write* W *for the Weyl group of* T *in* G *(Definition*

5.1). Then there is a finite-to-one correspondence from the discrete series of G onto the set of all regular metaplectic characters of $\tilde{T}$, modulo W. This correspondence may be described as follows. To each regular metaplectic character τ, one can associate a unitary representation $I(\tau) = I_T(\tau)$. We have

a) $I(\tau)$ is a direct sum of a finite number of discrete series representations.

b) Every discrete series representation of G occurs in some $I(\tau)$.

c) For w in W, $I(\tau)$ is equivalent to $I(w\tau)$.

d) If τ' is not in $W\tau$, then $I(\tau)$ and $I(\tau')$ have no irreducible constituents in common.

If G is in Harish-Chandra's class, then the various $I(\tau)$ are all irreducible, and the correspondence above is a bijection.

A few small points require attention. First, we began by hoping to attach representations of G to representations of compact Cartan subgroups. The parameter set turns out in the end to involve metaplectic representations. In the case of compact groups, we cured this problem by replacing the metaplectic representation τ of T by $\mu = \tau \otimes (-\rho)$. We cannot do that here, for a simple reason. We would like to recover τ from μ, as $\mu \otimes \rho$. The difficulty is that if μ is not regular, it does not define a set of positive roots uniquely; so we do not know which ρ to use. In the compact case this did not matter, for all possible choices were conjugate under the stabilizer of μ in W. In the non-compact case, W may be strictly smaller than the Weyl group of the root system. The result (already in $SL(2,\mathbb{R})$) is that the same μ is attached to several non-conjugate τ. There are ways around this problem, but none of them leads to as simple a parametrization as in Theorem 1.17. On the other hand, there are real advantages to using metaplectic parameters (in writing character formulas, for example – cf. Theorem 1.40). It therefore seems reasonable to use them throughout.

We turn now to the problem of constructing the discrete series representations of Theorem 5.11. Harish–Chandra's approach was based on generalizing Theorem 1.40. We will instead seek a version of Theorem 1.24. So fix $\mathfrak{b}$, T, and τ as in Theorem 5.11. The most obvious imitation of Theorem 1.24 would put on G/T the complex structure defined by (that is, having holomorphic tangent space corresponding to) $\mathfrak{b}$, and the line bundle $\mathscr{W}^-$ associated to the representation $\mu = \tau \otimes (-\rho)$ (Propositions 1.19 to 1.21). The space V of K-finite holomorphic sections of $\mathscr{W}^-$ is an admissible $(\mathfrak{g},K)$-

module. Unfortunately, it is the wrong one: the proof of Theorem 1.24 shows that it contains a copy of any finite-dimensional representation of G of highest weight μ. (Such a representation always exists if G is linear, since τ is assumed to be dominant and regular.) Finite-dimensional representations of non-compact reductive groups are rarely unitary and are never in the discrete series.

There is another obvious complex structure on G/T available, however: the one defined by $\mathfrak{b}^-$. Let (μ,W) denote the representation $\tau \otimes \rho$ of T. We make it into a $(\mathfrak{b},T)$-module by letting $\mathfrak{n}$ act by zero; then (Proposition 1.21) it defines a holomorphic vector bundle $\mathcal{W}$ on G/T. If G is compact and G_0 is non-abelian, Theorem 1.24 guarantees that this bundle has no holomorphic sections at all:

(5.13) $\Gamma(G/T, \mathcal{W}) = 0$.

It is not particularly difficult to prove this for general G as well. We need not despair, however. A holomorphic vector bundle has not only sections, but also higher cohomology groups.

Definition 5.14. Suppose $\mathcal{W}$ is a holomorphic vector bundle on a complex manifold X. Write $\mathcal{S}_{\mathcal{W}}$ for the sheaf of germs of holomorphic sections of X. The cohomology group $H^p(X, \mathcal{W})$ is by definition the pth sheaf cohomology group of X with coefficients in $\mathcal{G}_{\mathcal{W}}$. By Dolbeault's theorem, this is isomorphic to the cohomology defined using the $\bar{\partial}$ operator on $(0,p)$ forms with coefficients in $\mathcal{W}$.

The group G acts on the cohomology groups of a homogeneous holomorphic vector bundle; so we can look for the representations we want there. In case G is compact, Serre duality and Theorem 1.24 give

$$(5.14) \qquad H^S(K/T,\mathcal{W}) \cong \text{representation of}$$
$$\text{highest weight } \tau\otimes(-\rho),$$

which of course is exactly what we want for $I(\tau)$. Here S is the dimension of K/T as a complex manifold. For G non-compact, the cohomology groups are much more difficult to analyze. (For example, they are generally infinite-dimensional, and it is not clear that they carry Hausdorff topologies. To prove that, one must show that $\bar{\partial}$ operator has closed range.) Nevertheless, it is possible to study them.

THEOREM 5.15 ([Schmid, 1967] and [Schmid, 1975]). *Suppose G is a reductive Lie group, T is a large compact Cartan subgroup associated to a Borel subalgebra $\mathfrak{b}$, and τ is a* *dominant irreducible metaplectic representation of* $\tilde{T}$

(Definition 5.7). Assume (without loss of generality, by

Lemma 5.5) that T *is contained in* K. *Write* (μ, W) *for*

the representation $\tau \otimes \rho$ *of* T. *Endow* G/T *with the holo-*

morphic structure defined by $\mathfrak{b}^-$; *write* $\mathcal{W}$ *for the holomor-*

phic vector bundle defined by μ. *Let* S *be the complex*

dimension of K/T.

 a) The Harish-Chandra module of the discrete series

representation $I_{\mathfrak{b}, T}(\tau)$ *is isomorphic to the space of*

K-finite elements in $H^S(G/T, \mathcal{W})$.

 b) If i *is not equal to* S, *then* $H^i(G/T, \mathcal{W})$ *is zero.*

 c) K/T is a compact (complex) submanifold of maximal

dimension in G/T.

This is a reasonable generalization of the Borel-Weil theo-

rem (Theorem 1.24).

 There is a distinction to be made here among three

problems: *realizing* representations that are already known

to exist, as in Theorems 1.24 and 5.15; proving the *exis-*

tence of representations, as in Theorems 1.17 and 5.11; and

constructing representations (that is, doing existence and

realization at the same time), as in Example 3.20 and

Theorem 4.23. Clearly the third possibility is the most

desirable; the second without the first is in any case in-
adequate.

It is therefore a serious problem that the original

proof of Theorem 5.15 depends on knowing Theorem 5.11 in

advance. Although one can say quite a bit *a priori* about

the space V of K-finite elements in $H^S(G/T, \mathcal{W})$ from an

algebraic point of view, it is a difficult matter to impose

even an invariant pre-Hilbert space structure on V. To

some extent this problem exists already for compact groups.

Our proof of Theorem 1.24 invoked Theorem 1.17; eliminating

this dependence is possible but requires some moderately

sophisticated differential geometry. To approach Theorem

5.15 without Theorem 5.11 is far harder, but it is still

possible. To begin with, the Dolbeault cohomology should be

replaced by some kind of space of L^2 harmonic forms. (Of

course this is no change if G is compact, but the spaces

are very different in general.) Kostant and Langlands first

suggested that the discrete series could be realized on such

spaces. Their conjecture was proved in [Schmid, 1976],

still using Theorem 5.11. To make this into a construction

- that is, to prove *a priori* that the L^2 cohomology spaces

carry discrete series representations - is more or less the

point of [Atiyah-Schmid, 1977].

Unfortunately, we want even more unitary representations than are provided by Theorem 5.11. There is some hope
of finding these by generalizations of Theorem 5.15, and
this hope has been partially realized in [Rawnsley-Schmid-
Wolf, 1983]. The reader should consult that paper for more
information. For the present, however, the analytic difficulties are insuperable except in special cases.

The beginnings of a way out of these difficulties were
provided by Zuckerman in 1977 (see Chapter 6 of [Vogan,
1981]). In the setting of Theorem 5.15, one can think of
the idea in this way. The complex manifold G/T contains
K/T as a compact complex submanifold. Stein manifolds
cannot have non-trivial compact submanifolds; so G/T is
almost never Stein. Schmid showed, however, that G/T is
in a certain technical sense Stein away from K/T. Stein
manifolds have many holomorphic functions. One should therefore expect the only constraints on $H^S(G/T, \mathcal{W})$ to come from
K/T. More precisely, anything that looks like the jet of a
class in this cohomology near K/T ought to arise from a
genuine global cohomology class. Zuckerman's idea was to
set up a formalism to describe such jets and then to use
this formalism instead of the Dolbeault cohomology. The procedure gives only a Lie algebra representation, but Theorem
2.15 then immediately turns it into a group representation.

Most importantly, the procedure can be generalized substantially without much effort: we can build representations of
G out of representations of a wide class of Levi subgroups.
The next chapter is devoted to a description of Zuckerman's
construction.

Chapter 6

COHOMOLOGICAL PARABOLIC INDUCTION: ALGEBRAIC THEORY

We are going to study Zuckerman's construction in detail only under what may appear to be rather restrictive
hypotheses. We will therefore begin with a lemma intended
to motivate them.

LEMMA 6.1. *Suppose* L *is a Levi factor in* G, *associated
to a parabolic subalgebra* $q = l+u$ *in the complexification
(Definition 5.1). Assume that* L *is stable under the
Cartan involution* θ *(cf. Lemma 5.5). Then at least one of
the following possibilities holds.*

a) The centralizer of L *in* g_0 *(which is contained
in* l_0*) meets the* -1 *eigenspace* s_0 *of* θ*. In that case,*
L *is contained in the Levi factor of a proper real para-
bolic subgroup of* G.

b) The parabolic q *is* θ-*stable.*

123

This second condition is equivalent to

    b)' The parabolic $\mathfrak{q}$ *is opposite to its complex conju-gate* $\mathfrak{q}^-$. *That is,*

$$\mathfrak{q} \cap \mathfrak{q}^- = \mathfrak{l}.$$

Proof. Recall from Definition 5.7 the one-dimensional repre-sentation $2\rho(\mathfrak{u})$ of L. Since θ acts on L, it acts on representations of L. Clearly

$$\theta(2\rho(\mathfrak{u})) = 2\rho(\theta\mathfrak{u});$$

the Lie algebra $\theta\mathfrak{q}$ is another parabolic with Levi factor $\mathfrak{l}$. Define Φ to be the quotient of these two characters, and ϕ its differential in $\mathfrak{l}^*$. Then ϕ is zero if and only if $\mathfrak{q}$ is θ-stable. Furthermore ϕ is invariant under the coadjoint action of L, and $\theta\phi$ is $-\phi$. Elements of $\mathfrak{s}_0$ have real eigenvalues in the adjoint representation. Since ϕ is built from the adjoint representation, it fol-lows that ϕ is real valued. (The equivalence of (b) and (b)' is similar; we omit details.) Identifying $\mathfrak{l}_0$ with $(\mathfrak{l}_0)^*$ by the form chosen before (2.2), we conclude that ϕ corresponds to an L-fixed element of $\mathfrak{s}_0$. The proposition now follows from Lemma 5.8.   □

We would like to be able to construct unitary represen-tations of G out of unitary representations of Levi sub-

groups of G. If we are willing to do this in a step-by-step
manner, Lemma 6.1 says that it suffices to consider two special cases: Levi factors of real parabolic subgroups, and
Levi factors of θ-stable parabolic subalgebras. Since we
have treated the first case in Chapter 3, we will concentrate now on the second. For the rest of this chapter, we
will therefore work with the following situation.

$$q = \mathfrak{l}+u \quad \text{is a parabolic subalgebra of } \mathfrak{g}$$

$$\theta \text{ preserves } q, \ \mathfrak{l}, \text{ and } u$$

(6.2)(a) $\mathfrak{l} = $ complexification of $\mathfrak{l}_0$

$$L = \text{a Levi subgroup for } q.$$

Because θ preserves everything, the Cartan decomposition
(2.1) gives

$$q = q \cap \mathfrak{k} + q \cap \mathfrak{s}$$

(6.2)(b) $$u = u \cap \mathfrak{k} + u \cap \mathfrak{s}$$

$$L = (L \cap K) \cdot \exp(\mathfrak{l}_0 \cap \mathfrak{s}_0).$$

We write

(6.2)(c) $R = \dim u \cap \mathfrak{s}, \quad S = \dim u \cap \mathfrak{k}.$

The complex conjugate parabolic subalgebra q^- gives rise to
a triangular decomposition

(6.2)(d) $\mathfrak{g} = u + \mathfrak{l} + u^-.$

From Definition 5.7, we have the metaplectic cover $\tilde{L}$ of
L and its metaplectic character

(6.2)(e) $\rho(u) \in (\tilde{L})^{\wedge}_u.$

(The character is unitary for the following reason. By the proof of Lemma 6.1, its differential vanishes on $\mathfrak{l} \cap \mathfrak{s}$; so it takes purely imaginary values on $\mathfrak{g}_0$:

$$(6.2)(f) \qquad\qquad \rho(\mathfrak{u}) \in i(\mathfrak{l}_0 \cap \mathfrak{k}_0)^*.)$$

Because $L \cap K$ preserves the decompositions (6.2)(b), we have two more characters of $L \cap K$:

$$(6.2)(g) \qquad 2\rho(\mathfrak{u} \cap \mathfrak{k}) = \det(\mathrm{Ad}|_{\mathfrak{u} \cap \mathfrak{k}}),$$

$$\qquad\qquad = \text{character of } L \cap K \text{ on } \Lambda^S(\mathfrak{u} \cap \mathfrak{k}),$$

and (similarly) $2\rho(\mathfrak{u} \cap \mathfrak{s})$. Obviously the sum of these two characters is $2\rho(\mathfrak{u})$.

In order to describe even the most elementary proper-ties of Zuckerman's construction, we need a long digression on the center of the enveloping algebra.

Definition 6.3. Suppose $\mathfrak{g}$ is a complex reductive Lie algebra. Write

$$(6.3)(a) \qquad\qquad \mathcal{Z}(\mathfrak{g}) = \text{center of } U(\mathfrak{g}).$$

Fix a parabolic subalgebra and a Levi factor of it, $\mathfrak{q} = \mathfrak{l} + \mathfrak{u}$. Write $\mathfrak{u}^-$ for the nil radical of the opposite parabolic sub-algebra, so that

$$\mathfrak{g} = \mathfrak{u} + \mathfrak{l} + \mathfrak{u}^-.$$

By the Poincaré-Birkhoff-Witt theorem, the enveloping alge-bra decomposes as a vector space, as

$$U(\mathfrak{g}) = U(\mathfrak{u}) \otimes U(\mathfrak{l}) \otimes U(\mathfrak{u}^-).$$

In particular, there is a direct sum decomposition

(6.3)(b) $\qquad U(\mathfrak{g}) = U(\mathfrak{l}) \oplus [\mathfrak{u}U(\mathfrak{g}) + U(\mathfrak{g})\mathfrak{u}^-]$.

(The sum within the square brackets is not direct.) Write

(6.3)(c) $\qquad\qquad \tilde{\xi} : U(\mathfrak{g}) \to U(\mathfrak{l})$

for the projection on the first factor. Let Z be any element of the center of $\mathfrak{l}$ such that $\mathrm{ad}(Z)$ has positive eigenvalues on $\mathfrak{u}$. Then the restriction of $\tilde{\xi}$ to the centralizer of Z is a homomorphism of algebras.

Recall the one-dimensional representation $\rho(\mathfrak{u})$ of $\mathfrak{l}$ (Definition 5.7). Define a map T_q from $\mathfrak{l}$ into $U(\mathfrak{l})$ by

(6.3)(d) $\qquad\qquad T_q(X) = X + \rho(\mathfrak{u})(X)$.

Because $\rho(\mathfrak{u})$ is a Lie algebra homomorphism, so is T_q. It therefore extends uniquely to a homomorphism of algebras

$$T_q : U(\mathfrak{l}) \to U(\mathfrak{l}).$$

Because it has an obvious inverse, T_q is actually an automorphism. The *Harish-Chandra map* is

(6.3)(e) $\qquad\qquad \xi = \xi_q = T_q \circ \tilde{\xi}$,

a map from $U(\mathfrak{g})$ to $U(\mathfrak{l})$. This map respects the adjoint action of the group L:

(6.3)(f) $\quad \xi[\mathrm{Ad}_{\mathfrak{g}}(x)(u)] = \mathrm{Ad}_{\mathfrak{l}}(x)(\xi(u)) \qquad (x{\in}L,\ u{\in}U(\mathfrak{g}))$.

Restricted to the centralizer of any element Z as above, it is a homomorphism of algebras. In particular, we have the *Harish-Chandra homomorphism*

(6.3)(g) $\qquad\qquad \xi : \mathcal{Z}(\mathfrak{g}) \to \mathcal{Z}(\mathfrak{l})$.

Suppose G is a reductive group with complexified Lie

algebra $\mathfrak{g}$. Write

(6.3)(h) $\mathcal{Z}_G(\mathfrak{g})$ = Ad(G)-invariants in $U(\mathfrak{g})$ $\subseteq$ $\mathcal{Z}(\mathfrak{g})$.

The containment is an equality if and only if G belongs to

Harish-Chandra's class. Because of (6.3)(f), we can re-

strict ξ to

(6.3)(i) ξ: $\mathcal{Z}_G(\mathfrak{g})$ → $\mathcal{Z}_L(\mathfrak{l})$.

THEOREM 6.4 (Harish-Chandra; cf. [Humphreys, 1972]). *Sup-*

pose we are in the setting of Definition 6.3.

   a) *The Harish-Chandra homomorphism* ξ *of (6.3)(g)*

depends only on $\mathfrak{l}$, *and not on* $\mathfrak{q}$. *Consequently, we can*

write

$$\xi_{\mathfrak{l}} \colon \mathcal{Z}(\mathfrak{g}) \to Z(\mathfrak{l})^{W(\mathfrak{g},\mathfrak{l})}$$

(Definition 5.1).

   b) $\xi_{\mathfrak{l}}$ *is injective.*

   c) *If* $\mathfrak{l} = \mathfrak{h}$ *is a Cartan subalgebra, then*

$$\xi_{\mathfrak{h}} \colon \mathcal{Z}(\mathfrak{g}) \to S(\mathfrak{h})^{W(\mathfrak{g},\mathfrak{l})}$$

is an isomorphism. In the setting of (6.3)(i), write $\widetilde{W}$

for the group of automorphisms of $\mathcal{Z}_L(\mathfrak{l})$ *generated by*

$W(G,L)$ *and* $W(\mathfrak{g},\mathfrak{l})$. *Then we have*

   d) *The map*

$$\xi_L \colon \mathcal{Z}_G(\mathfrak{g}) \to \mathcal{Z}_L(\mathfrak{l})^{\widetilde{W}},$$

is injective.

e) *Suppose that* $L = H$ *is a Cartan subgroup and that* $W(G,H)$ *has a representative in each connected component of* G. (*This is automatic if* H *is small and is either fundamental or maximally split.*) *Then* ξ_L *is an isomorphism.*

Definition 6.5. We use the notation of Definition 6.3. Fix a Cartan subalgebra $\mathfrak{h}$ of $\mathfrak{g}$. Each element λ of $\mathfrak{h}^*$ defines a ring homomorphism

$$\chi_\lambda : S(\mathfrak{h}) \to \mathbb{C},$$

by evaluation at λ. This gives rise to

$$\xi_\lambda = \xi_{\mathfrak{h},\lambda} = \chi_\lambda \circ \xi_{\mathfrak{h}},$$

a homomorphism from $\mathcal{Z}(\mathfrak{g})$ to $\mathbb{C}$. We call such a homomorphism an *infinitesimal character*. Equivalently, there is a maximal ideal

$$\mathcal{J}_\lambda = \ker \xi_\lambda$$

in $\mathcal{Z}(\mathfrak{g})$. Again Theorem 6.4 makes $\xi_{\mathfrak{h}}$ an integral ring extension; so all maximal ideals in $\mathcal{Z}(\mathfrak{g})$ arise in this way. We write

$$\text{Max } \mathcal{Z}(\mathfrak{g}) = \mathfrak{h}^*/W(\mathfrak{g},\mathfrak{h}).$$

A similar discussion applies to $\mathcal{Z}_G(\mathfrak{g})$.

More generally, suppose $\mathfrak{l}$ is a Levi subalgebra of $\mathfrak{g}$, and ψ is an infinitesimal character for $\mathfrak{l}$. Define

$$\xi_\psi = \xi_{\mathfrak{l},\psi} = \psi \circ \xi_{\mathfrak{l}},$$

a homomorphism from $\mathcal{Z}(\mathfrak{g})$ to $\mathbb{C}$.

We say that a $\mathfrak{g}$ module V *has infinitesimal charac-*
ter λ (or ξ_λ) if V is annihilated by $\mathcal{I}_\lambda$:

$$z \cdot v = \xi_\lambda(z) \cdot \qquad (z \in \mathcal{Z}(\mathfrak{g}), \ v \in V).$$

Again, a similar definition applies to $\mathcal{Z}_G(\mathfrak{g})$. We will say
G-*infinitesimal character* if it is necessary to emphasize
the distinction.

LEMMA 6.6 (algebraic Schur's lemma; see [Dixmier, 1974]).
Any irreducible $\mathfrak{g}$ *module has an infinitesimal character;*
that is, $\mathcal{Z}(\mathfrak{g})$ *acts by scalars on it. Any irreducible*
$(\mathfrak{g},K)$-*module has a* G-*infinitesimal character.*

We could now begin to describe Zuckerman's construction, but a final remark about the case of real parabolics
may be instructive.

PROPOSITION 6.7. *In the setting of Definition 4.1, suppose*
that the representation ϕ *of* L *has* L-*infinitesimal*
character ψ *(Definition 6.5). Then the representation*
$\text{Ind}_P^G(\phi)$ *has* G-*infinitesimal character* $\xi_{L,\psi}$ *(Definition*
6.5).

This is proved from the definitions by a very simple calculation; details may be found in [Wallach, 1973] or [Knapp,
1986], for example.

THEOREM 6.8 (Zuckerman, Vogan; see [Vogan, 1981] and [Vogan,
1984]). *Suppose* L *is a Levi subgroup of the reductive
group* G, *attached to the* θ-stable parabolic subalgebra
q = l+u. *Use the notation* (6.2); *in particular,* L~ *is the
metaplectic double cover of* L, *and* S *is the dimension of*
u ∩ k. *Then there is a family*

$$\mathcal{R}^j = (\mathcal{R}_{q,L})^j \qquad (j=0,1,\ldots,S)$$

of functors from the category of metaplectic (l,(L ∩ K)~)-
modules, to the category of (g,K)-*modules, with the following properties. Let* Z *be an* (l,(L ∩ K)~)-*module.*

a) $\mathcal{R}^*$ *takes short exact sequences of metaplectic*
(l,(L ∩ K)~)-*modules to long exact sequences of* (g,K)-
modules.

b) *If* Z *has finite length, then so do all the* $\mathcal{R}^j Z$.
Let $\mathfrak{h}$ *be a Cartan subalgebra of* l. *Assume from now on
that* Z *has* L-*infinitesimal character* λ *in* $\mathfrak{h}^*$ *(Definition* 6.5).

c) $\mathcal{R}^j(Z)$ *has* G-*infinitesimal character* λ.

d) *Assume that for each root* α *of* $\mathfrak{h}$ *in* u ,

$$Re\langle\lambda,\alpha\rangle \geq 0.$$

Then $\mathscr{R}^j(Z)$ is zero for j not equal to S.

e) Under the hypothesis of (d), any non-degenerate Hermitian form on Z induces one on $\mathscr{R}^S(Z)$. If the former is definite, so is the latter. That is, if Z is infinitesimally unitary, then so is $\mathscr{R}^S(Z)$.

f) Assume that for each root α of $\mathfrak{h}$ in $\mathfrak{u}$,

$$Re\langle\lambda,\alpha\rangle > 0.$$

Then if Z is non-zero, so is $\mathscr{R}^S(Z)$.

g) Suppose that $L = T$ is a large compact Cartan subgroup of G (so that $\mathfrak{q} = \mathfrak{b}$ is a Borel subalgebra), and that τ is an irreducible dominant regular metaplectic representation of T. Then Harish-Chandra's discrete series is

$$I_{\mathfrak{b},T}(\tau) = (\mathscr{R}_{\mathfrak{b},T})^S(\tau).$$

(The only result not due to Zuckerman here is the statement about definiteness in (e).)

We call the functors $\mathscr{R}^j$ cohomological parabolic induction. Because of its special importance, we will use a special notation for $\mathscr{R}^S$:

(6.9) $I_{\mathfrak{q},L} = (\mathscr{R}_{\mathfrak{q},L})^S.$

The reader should be warned that the functor $\mathscr{R}^S$ considered here differs by tensoring with $\rho(\mathfrak{u})$ from the one defined in [Vogan, 1981]. Letters of complaint on this point may be

addressed to M. Duflo, who crystallized for me some previously half-formed ideas about it.

Notice that the "dominant regular" condition in (g) is
the same as the condition in (f). The rest of the theorem
therefore guarantees the existence of the proposed discrete
series as unitary representations. (This is circular unless
one uses the proof in [Wallach, 1984] for (e).) It is fair-
ly easy to show that $(\mathcal{R}_{\mathfrak{b},T})^S(\tau)$ is irreducible. To show
that it belongs to the discrete series requires some of
Harish-Chandra's basic ideas about the Schwartz space of G,
but far less serious analysis than is needed for Harish-
Chandra's proof of Theorem 5.11. To show that every discrete series is of this form is still more difficult, but
can be done using some of the ideas of Langlands and others
discussed in Chapter 13.

The philosophical content of Theorem 6.8 is this.
According to Proposition 1.19, $\mathfrak{q}$ defines a complex structure on G/L (cf. Lemma 6.1(b)'). Write

(6.10) $$W = Z \otimes \mathbb{C}_{\rho(\mathfrak{u})}.$$

Since Z and $\rho(\mathfrak{u})$ are both metaplectic, W descends to
an $(\mathfrak{l},(L \cap K))$-module. Asume that W is the Harish-Chandra
module for a representation $(\omega,\mathcal{H}_\omega)$ of L. After replacing
$\mathcal{H}_\omega$ by the subspace of smooth vectors, we may assume that $\mathfrak{l}$

acts on it. Extend ω to a module for $\mathfrak{q}$ by making $\mathfrak{u}$

act trivially. Then ω defines a (possibly infinite-

dimensional) holomorphic vector bundle $\mathcal{W}$ on G/L (Propo-

sition 1.21). Definition 5.14 therefore provides sheaf

cohomology groups, on which G acts. It is conjectured

that

(6.11) $(\mathcal{R}_{\mathfrak{q},L})^j(Z) \cong$ Harish-Chandra module of $H^j(G/L,\mathcal{W})$.

Theorem 5.15 (in conjunction with Theorem 6.8) is a special

case of this conjecture.

Conjecture (6.11) sheds some light on Theorem 6.8.

Most obviously, it explains why we should get a family of

functors instead of just one. As an analogue of parabolic

induction, the functor $\mathcal{R}^S$ has as its main flaw the strong

requirement in Theorem 6.8(e) needed to guarantee unitarity.

This is a completely unavoidable problem, however. The dis-

cussion after Theorem 5.12 included the observation that

even if L is compact, $H^0(G/L,\mathcal{W})$ may contain a (non-

unitary) finite-dimensional representation. A slightly more

sophisticated argument will produce non-unitary finite-

dimensional representations inside $H^S(G/L,\mathcal{W})$ as well.

Conjecture 6.11 then suggests that $\mathcal{R}^S$ does not preserve

unitarity in these cases, and in fact one can prove that

very easily. A second unsatisfactory point is that $\mathcal{R}^S$

depends on $\mathfrak{q}$, and not just on L (compare Theorem 3.19).

The same examples show that this is definitely the case,
however: in the setting of Theorem 6.8(g), we can sometimes
find a τ and two different Borel subalgebras, such that
one of the functors produces an irreducible discrete series
representation, and the other something containing a finite-
dimensional representation. If τ is allowed to be singular, we can sometimes choose the two Borel subalgebras both
to satisfy the condition in (d) of the theorem, and get two
different unitary representations of G.

We turn now to the construction of the functors $\mathcal{R}^j$.

Definition 6.12. Suppose $\mathfrak{g}$ is a Lie algebra and $\mathfrak{h}$ is a
subalgebra. Let V be any $\mathfrak{h}$ module. The *produced module*
from $\mathfrak{h}$ to $\mathfrak{g}$ is

$(6.12)(a)$ $\mathrm{pro}_{\mathfrak{h}}^{\mathfrak{g}}(V) = \mathrm{Hom}_{\mathfrak{h}}(U(\mathfrak{g}),V)$.

The Hom is defined using the left action of $\mathfrak{h}$ on $U(\mathfrak{g})$
and made into a $\mathfrak{g}$ module using the right action. More
precisely, a linear map ϕ from $U(\mathfrak{g})$ to V must satisfy

$\phi(Zu) = Z \cdot (\phi(u))$ $(Z \in \mathfrak{h}, \ u \in U(\mathfrak{g}))$

in order to belong to the produced module. In that case, an
element X in $\mathfrak{g}$ acts by

$(X \cdot \phi)(u) = \phi(uX)$.

The *order of vanishing filtration* of $\mathrm{pro}(V)$ is the decreasing filtration defined by

(6.12)(b) $[pro(V)]_n = \{\phi \in pro(V) \mid \phi(u) = 0, \text{ all } u \in U_n(\mathfrak{g})\}$

Here $U_n(\mathfrak{g})$ is the subspace of $U(\mathfrak{g})$ spanned by products of at most n elements of $\mathfrak{g}$. Notice that $pro(V)$ is complete for this filtration.

Suppose that M is a Lie group and we are given the ingredients of Definition 1.26 for the pair $(\mathfrak{g},M)$. Assume in addition that $\mathfrak{h}$ contains $\mathfrak{m}$ and that $Ad(M)$ preserves $\mathfrak{h}$; then $(\mathfrak{h},M)$ also satisfies the requirements of Definition 1.26. Suppose now that V is an $(\mathfrak{h},M)$-module. We can define an action of M on $pro_{\mathfrak{h}}^{\mathfrak{g}}(V)$ by

$$(b \cdot \phi)(u) = b \cdot (\phi(Ad(b^{-1})u)).$$

This action is usually not locally finite (Definition 1.26 (a)), but we can certainly consider the largest subspace where it is locally finite. In analogy with Definition 2.4, we denote this by a subscript M and define

(6.12)(b) $pro_{(\mathfrak{h},M)}^{(\mathfrak{g},M)}(V) = Hom_{\mathfrak{h}}(U(\mathfrak{g}),V)_M.$

By way of motivation, we return first to the C^∞ category.

PROPOSITION 6.13 (invariant Taylor series). *Suppose* G *is a Lie group,* H *is a closed subgroup,* $\mathcal{V}$ *is a finite-dimensional homogeneous vector bundle on* G/H, *and* (π,V) *is the corresponding representation of* H *(Propositions*

1.20 and 3.2). Write $C^\infty(G/H, \mathscr{V})$ for the space of smooth sections of $\mathscr{V}$ and X for the subspace of sections vanishing to infinite order at eH. Then there is a short exact sequence of representations of $\mathfrak{g}$ and of H

$$0 \to X \to C^\infty(G/H, \mathscr{V}) \to \mathrm{pro}_\mathfrak{h}^\mathfrak{g}(V) \to 0.$$

The second map preserves the order of vanishing filtrations.

Proof. Write λ for the action of $U(\mathfrak{g})$ on smooth functions by infinitesimal left translation. For X in $\mathfrak{g}_0$, this is given by

$$(6.14) \qquad (\lambda(X)f)(g) = (d/dt)(f(\exp(-tX)g))\big|_{t=0}.$$

The action is extended to $U(\mathfrak{g})$ so as to give an algebra representation. Suppose then that F is a smooth section of $\mathscr{V}$. Identify it with a function f from G to V, by Corollary 3.4. Define a map ϕ_F from $U(\mathfrak{g})$ to V by

$$(6.15) \qquad \phi_F(u) = (\lambda(u)f)(e).$$

That this map belongs to the produced module, and has the required properties, follows from Corollary 3.4 and elementary arguments. (One needs to know at some point that a smooth function can have any prescribed Taylor series, but this is a standard fact from analysis.) We omit the details. $\square$

COROLLARY 6.16. *In the setting of Proposition 6.13, assume that G/H is connected. Then the space* $C^\omega(G/H, \mathcal{V})$ *of real-analytic sections of* $\mathcal{V}$ *embeds in* $\mathrm{pro}_\mathfrak{g}^\mathfrak{h}(V)$. *More precisely (even if G/H is not connected), the space of germs at eH of real analytic sections embeds in* $\mathrm{pro}(V)$.

Next, we introduce a complex structure.

PROPOSITION 6.17. *In the setting of Proposition 6.13, suppose G/H has an invariant complex structure given by* $\mathfrak{b}^-$ *(Proposition 1.19), and that* $\mathcal{V}$ *has a holomorphic structure given by a* $(\mathfrak{b}, H)$*-module structure on* V. *Write* $\mathcal{S}$ *for the sheaf of germs of holomorphic sections of* $\mathcal{V}$. *Then there is a natural inclusion*

$$\mathcal{S}_{eH} \rightarrow \mathrm{pro}_\mathfrak{b}^\mathfrak{g}(V),$$

preserving the actions of $U(\mathfrak{g})$ *and the order of vanishing filtrations. If* $\mathcal{S}_{eH}$ *is completed with respect to this filtration, the map becomes an isomorphism.*

This is clear from Proposition 1.21 and Corollary 6.16. The completion of a space of germs of holomorphic sections with respect to the order of vanishing filtration is just the space of formal power series sections.

From this point on, we will be considering some more
subtle operations on $(\mathfrak{g},K)$-modules. Some of these cannot
be carried out, or require much more care, in the general
context of Definition 1.26. It is therefore convenient to
introduce a more restrictive version of that definition.

Definition 6.18. Suppose $(\mathfrak{h},M)$ is a pair satisfying the
hypotheses (i) and (ii) imposed on $(\mathfrak{b},H)$ in Definition
1.26. Assume in addition that

   iii) M is compact.

Suppose X is an $(\mathfrak{h},M)$-module. Write X^d for the
M-finite vectors in the dual space of X. In notation like
that of Definition 6.12,

   a) $X^d = \mathrm{Hom}_{\mathbb{C}}(X,\mathbb{C})_M .$

This is again an $(\mathfrak{h},M)$-module, and the functor sending X
to X^d is exact. We call X^d the *M-finite dual* of X.

   Similarly, let X^h denote the $(\mathfrak{h},M)$-module whose
underlying real vector space is X^d, but whose multiplication by i (the square root of -1) is the negative of that
for X^d. The actions of $\mathfrak{h}$ and M on X^h are unchanged
from the actions on X^d. There is a non-degenerate sesqu-
linear pairing between X and X^h. We call X^h the *Hermitian dual* of X. An $(\mathfrak{h},M)$-invariant sesquilinear pairing

between X and another ($\mathfrak{h}$,M) module Y amounts to a map from Y to X^h.

The assumption that M be compact simplifies the proof that the category of ($\mathfrak{h}$,M)-modules has enough injectives and allows one to prove that there are enough projectives. One can therefore speak of

b) $\mathrm{Ext}^i_{(\mathfrak{h},M)}$

(the derived functors of Hom). In particular, we define

c) $H^i(\mathfrak{h},M,X) = \mathrm{Ext}^i_{(\mathfrak{h},M)}(\mathbb{C},X)$,

the *relative Lie algebra cohomology groups*.

A detailed discussion of the category of ($\mathfrak{h}$,M) mod-ules and its Ext functors may be found in [Borel-Wallach, 1980] or [Vogan, 1981].

Definition 6.19 (Zuckerman). Suppose K is a Lie group, $\mathfrak{g}$ is a Lie algebra, and we are given the ingredients of Defini-tion 6.18 for the pair ($\mathfrak{g}$,K). Let M be a closed subgroup of K; then we have these ingredients for ($\mathfrak{g}$,M) as well. We define a functor (the K-*finite vectors in* functor),

$$\Gamma = \Gamma^{(\mathfrak{g},K)}_{(\mathfrak{g},M)}$$

from ($\mathfrak{g}$,M) modules to ($\mathfrak{g}$,K) modules, as follows. Let X be a ($\mathfrak{g}$,M) module. Set

$$(\Gamma_0)^{\sim}(X) = \{v \in X| \dim U(\mathfrak{k})v < \infty\},$$

the space of $\mathfrak{k}$-finite vectors. By elementary Lie group theory, this space carries a representation of the universal cover $(K_0)^{\sim}$ of the identity component of K. Write Z for the kernel of the covering map, and

$$\Gamma_0(X) = \{v \in (\Gamma_0)^{\sim}(X)| z \cdot v = v, \text{ all } z \in Z\}.$$

Obviously $\Gamma_0(X)$ carries a representation of K_0. Call this representation π for a moment (although we will soon return to module notation), and write ρ for the representation of M on X. Set

$$\Gamma_1(X) = \{v \in \Gamma_0(X)| \pi(m)v = \rho(m)v, \text{ all } m \in M \cap K_0\}.$$

Let K_1 be the subgroup of K generated by K_0 and M. Then there is a unique action of K_1 on $\Gamma_1(X)$, extending both π and ρ. Finally, set

$$\Gamma(X) = \text{Ind}_{K_1}^{K}(\Gamma_1(X)),$$

with induction defined formally as in Definition 3.8. (The hypotheses of that definition are not satisfied here, but that only means we cannot get a Hilbert space.) We make $\mathfrak{g}$ act on the induced representation by

$$(X \cdot f)(k) = (\text{Ad}(k^{-1})(X)) \cdot (f(k)).$$

Although this definition requires care, it should not be re-garded too seriously; all we want is (to make sense of) the "subspace" of X on which the action of $\mathfrak{k}$ exponentiates

to K. If K and M are connected, and K is simply connected, then $\Gamma(X)$ is the (easily defined) subspace of
$\mathfrak{k}$-finite vectors in X. One should keep this case in mind.

Definition 6.20 (Zuckerman; see [Vogan, 1981]). We use the
notation of (6.2). Suppose Z is a metaplectic $(\mathfrak{l},(L\cap K)^{\sim})$-
module. Write
$$W = Z \otimes \mathbb{C}_{\rho(\mathfrak{u})}$$
(compare (6.10)). By (6.2)(e), W is an $(\mathfrak{l},(L\cap K))$-module.
Extend W to a $(\mathfrak{q},(L\cap K))$-module by making $\mathfrak{u}$ act trivially. Define (following Definition 6.12)
$$X = \text{pro}_{(\mathfrak{g},L\cap K)}^{(\mathfrak{g},L\cap K)}(W).$$
We define (following Definition 6.19)

(6.20)(a) $(\mathcal{R}_{\mathfrak{q},L})^{0}(Z) = \Gamma_{(\mathfrak{g},L\cap K)}^{(\mathfrak{g},K)}(X).$

It is a simple matter to verify that $\mathcal{R}^{0}$ is a left exact
functor. Because the category of $(\mathfrak{l},(L\cap K)^{\sim})$-modules has
enough injectives, we can define

(6.20)(b) $(\mathcal{R}_{\mathfrak{q},L})^{i} = i^{\text{th}}$ right derived functor of $\mathcal{R}^{0}$.

In fact all the functors involved in the definition of $\mathcal{R}$
are exact and take injectives to injectives, except for Γ.
We could therefore define $\mathcal{R}^{i}$ by replacing Γ by its ith
right derived functor in (6.20)(a).

Here are some comments about the proof of Theorem 6.8.
Part (a) is elementary homological algebra (except that we
have not yet explained why $\mathcal{R}^i$ is zero for i greater than
S). Consider next (c). Then obviously W (Definition
6.20) has infinitesimal character $\lambda + \rho(u)$. We now apply
the following infinitesimal version of Proposition 6.7.

LEMMA 6.21. *In the setting of Definition 6.3, suppose* W
is a module for $\mathfrak{l}$ *, extended to* $\mathfrak{q}$ *by making* u *act triv-*
ially. Suppose z *is in* $\mathcal{Z}(\mathfrak{g})$*, and* $\tilde{\xi}$ *is the map of*
(6.3)(c). Then $\tilde{\xi}(z)$ *is in* $\mathcal{Z}(\mathfrak{l})$ *and so defines a* $\mathfrak{q}$-
module map ϕ *from* W *to* W*. Write* Φ *for the* $\mathfrak{g}$ *module*
map induced (by functoriality) on $\mathrm{pro}_{\mathfrak{q}}^{\mathfrak{g}}(W)$ *(Definition*
6.12). Then the action of z *on* $\mathrm{pro}(W)$ *is given by* Φ*.*

In particular, if W *has infinitesimal character*
$\lambda + \rho(u)$*, then* $\mathrm{pro}(W)$ *has infinitesimal character* λ *(Defi-*
nition 6.5).

It is easy to see that the functor Γ preserves G-infinite-
simal character. Lemma 6.21 therefore leads to

COROLLARY 6.22. *In the setting of Definition 6.20, suppose*
z *is in* $\mathcal{Z}_G(\mathfrak{g})$*. Write* $\xi(z)$ *for its image in* $\mathcal{Z}_L(\mathfrak{l})$ *un-*

der the Harish-Chandra map (6.3)(i). Write ϕ for the action of $\xi(z)$ on Z, and Φ^i for the map induced (by functoriality) on $\mathcal{R}^i(Z)$. Then Φ^i is the action of z on $\mathcal{R}^i$.

This includes Theorem 6.8(c).

To continue the proof of Theorem 6.8, it is necessary to analyze the restriction of $\mathcal{R}^i$ to K. The only diffi-culty is in understanding the derived functors of Γ. For that, we begin with an easy lemma.

LEMMA 6.23. *In the setting of Definition 6.19, suppose* X *is a* (g,M)-*module. Write* $\mathcal{F}X$ *for the underlying* $(\mathfrak{k},M)$-*module. Then*

$$\mathcal{F}\Gamma^{(g,K)}_{(g,M)}(X) = \Gamma^{(\mathfrak{k},K)}_{(\mathfrak{k},M)}(\mathcal{F}X).$$

(Here we have used $\mathcal{F}$ *on the left to denote the underlying* $(\mathfrak{k},K)$-*module for a* (g,K)-*module.) The same result holds for the derived functors.*

Proof. *The result for* Γ *follows by inspection of Defini-tion 6.19. For the derived functors, it suffices to show that there are enough injective* (g,M)-modules that are injective as* $(\mathfrak{k},M)$-modules. To see this, let* Y *be any locally finite representation of* M. *Then*

$$I = \text{Hom}_m(U(\mathfrak{g}),Y)_M$$

(Definition 6.12) is an injective $(\mathfrak{g},M)$-module, and there are enough injectives of this kind ([Vogan, 1981], Corollary 6.1.24). But

$$\mathscr{F}I = \text{Hom}_m(U(\mathfrak{k}),\text{Hom}(S(\mathfrak{g}/\mathfrak{k}),Y))_M$$

is an injective $(\mathfrak{k},M)$-module.   □.

Another basic fact is that Γ is an adjoint to another kind of forgetful functor.

LEMMA 6.24 (Zuckerman; see [Vogan, 1981], Lemma 6.2.10). *In the setting of Definition 6.19, suppose* V *is a* $(\mathfrak{g},K)$-*module. Write* $\mathscr{F}X$ *for the underlying* $(\mathfrak{g},M)$-*module. Then there is a natural isomorphism*

$$\text{Hom}_{(\mathfrak{g},K)}(V,\Gamma X) \cong \text{Hom}_{(\mathfrak{g},M)}(\mathscr{F}V,X).$$

This is fairly clear from the definition of Γ. An immediate consequence (using Lemma 6.23) is a corresponding statement about derived functors.

PROPOSITION 6.25. *In the setting of Definition 6.19, suppose* V *is a (locally finite) representation of* K. *Then there are natural isomorphisms*

$$\operatorname{Hom}_K(V, \Gamma^j X) \cong \operatorname{Ext}^j_{(\mathfrak{k},M)}(V,X)$$
$$\cong \operatorname{Ext}^j_{(\mathfrak{k},M)}(\mathbb{C}, \operatorname{Hom}_{\mathbb{C}}(V,X)_M)$$
$$\cong H^j(\mathfrak{k}, M, \operatorname{Hom}_{\mathbb{C}}(V,X)_M).$$

(We have omitted various forgetful functors.)

This proposition is the tool that allows one to compute the functors $\mathcal{R}^j$ as representations of K. We will return to that task in a moment. First, however, we study Hermitian forms.

LEMMA 6.26 (Poincaré duality; see [Borel-Wallach, 1980] or [Knapp-Vogan, 1986]). *In the setting of Definition 6.18, suppose* X *is an* $(\mathfrak{h}, M)$-*module. Write* t *for the dimension of* $\mathfrak{h}/\mathfrak{m}$. *Then*

$$\mathbb{C}_t = \Lambda^t(\mathfrak{h}/\mathfrak{m})$$

is a one-dimensional representation of M *in a natural way. We make* $\mathbb{C}_t$ *into an* $(\mathfrak{h}, M)$-*module by making an element* H *of* $\mathfrak{h}$ *act by* $\det(\operatorname{ad}(H))$; *that is, by the scalar by which it acts on the top exterior power of* $\mathfrak{h}$.

Then there is a natural isomorphism

$$H^i(\mathfrak{h}, M, X)^* \cong H^{t-i}(\mathfrak{h}, M, X^d \otimes \mathbb{C}_t).$$

Sketch of proof. The cohomology is computed by a certain complex

$$(6.27) \qquad \text{Hom}_M(\Lambda^i(\mathfrak{h}/\mathfrak{m}), X)$$

(see for example [Borel-Wallach, 1980]). Exterior multiplication defines an isomorphism

$$\Lambda^{t-i}(\mathfrak{h}/\mathfrak{m}) \cong \text{Hom}(\Lambda^i(\mathfrak{h}/\mathfrak{m}), \Lambda^t(\mathfrak{h}/\mathfrak{m})).$$

Evidently the dual of the complex (6.27) may therefore be identified with

$$\text{Hom}_M(\Lambda^{t-i}(\mathfrak{h}/\mathfrak{m}), X^d \otimes \Lambda^t(\mathfrak{h}/\mathfrak{m}))$$

One can check that in this identification, the transpose of the differential for X is the differential for $X^d \otimes \mathbb{C}_t$. It follows that

$$H^i(\mathfrak{h}, M, X)^* \cong H^{t-i}(\mathfrak{h}, M, X^d \otimes \mathbb{C}_t). \qquad \square.$$

THEOREM 6.28 ([Enright-Wallach, 1980]). *In the setting of Definition 6.19, let* X *be a* $(\mathfrak{g}, M)$*-module. Write* t *for the dimension of* $\mathfrak{k}/\mathfrak{m}$. *Then*

$$\mathbb{C}_t = \Lambda^t(\mathfrak{k}/\mathfrak{m})$$

is a one-dimensional representation of M *in a natural way. Make* $\mathbb{C}_t$ *into a* $(\mathfrak{g}, M)$*-module by making* $\mathfrak{g}$ *act trivially. Then there is a natural isomorphism of* $(\mathfrak{g}, K)$*-modules*

$$(\Gamma^i X)^h \cong \Gamma^{t-i}(X^h \otimes \mathbb{C}_t).$$

In particular, suppose that M *acts trivially on* $\Lambda^t(\mathfrak{k}/\mathfrak{m})$.

Then a (non-degenerate) invariant Hermitian form on X
induces a (non-degenerate) invariant Hermitian form on
$\Gamma^{t/2}(X)$.

Sketch of proof. Let V be any finite-dimensional representation of K. A calculation using Proposition 6.25 and
Lemma 6.26 shows that

$$\text{Hom}_K(V,\Gamma^i(X)^h) \cong \text{Hom}_K(V,\Gamma^{t-i}(X^h \otimes \mathbb{C}_t))$$

in a natural way. It follows that the desired isomorphism
exists on the level of representations of K. Enright and
Wallach now use a clever formal argument to deduce that a
sufficiently nice isomorphism as representations of K must
automatically respect the action of G as well; we refer to
their paper for the details.   □

PROPOSITION 6.29 (Shapovalov). *In the setting of Definition
6.20, an invariant Hermitian form on* Z *induces one on*

$$X^{\sim} = \text{ind}_{(q^-,L\cap K)}^{(g,L\cap K)} (Z \otimes \mathbb{C}_{\rho(u)}).$$

*Under the hypothesis of Theorem 6.8(d), the induced form is
non-degenerate, and* $X^{\sim}$ *is isomorphic to* X.

This is more or less routine; details may be found in
[Vogan, 1984]. (Shapovalov proved much deeper results about
the form on the induced module.)

We now return to the analysis of the K-types of $\mathscr{R}^j(Z)$.

Definition 6.30. In the setting of Definition 6.20, suppose $Z_{L\cap K}$ is a metaplectic $(q \cap \mathfrak{k}, (L \cap K)^{\sim})$-module. Set

$$W_{L\cap K} = Z_{L\cap K} \otimes \mathbb{C}_{\rho(u)}$$
$$X_{\mathfrak{k}} = \text{pro}((q \cap \mathfrak{k}, L \cap K) \uparrow (\mathfrak{k}, K))(W_{L\cap K})$$
$$(\mathscr{R}_{q\cap\mathfrak{k},L\cap K})^0(Z_{L\cap K}) = \Gamma^{(k,K)}_{(k,L\cap K)}(X_{\mathfrak{k}}).$$
$$(\mathscr{R}_{q\cap\mathfrak{k},L\cap K})^i = i^{\text{th}} \text{ right derived functor of } \mathscr{R}^0.$$

These are essentially taken from Definition 6.20, applied to K instead of G. The only difference is that we do not require $u \cap \mathfrak{k}$ to act trivially on $Z_{L\cap K}$. As in the earlier definition, we could simply take derived functors of Γ.

PROPOSITION 6.31. *In the setting of Definition 6.30, suppose $Z_{L\cap K}$ is finite-dimensional. Let $\mathscr{W}_{L\cap K}$ be the holomorphic vector bundle on $K/(L \cap K)$ associated to $W_{L\cap K}$; we have endowed the base space with the complex structure given by $(q \cap \mathfrak{k})^-$. Then there is a natural isomorphism*

$$(\mathscr{R}_{q\cap\mathfrak{k},L\cap K})(Z_{L\cap K}) \cong H^i(K/(L \cap K), \mathscr{W}_{L\cap K})$$

(Definition 5.14).

This is another special case of the conjecture (6.11); it is due to Zuckerman (unpublished). Its importance was first emphasized in [Enright-Wallach, 1980], where one can find a proof. The idea is that both sides of the equality in the proposition are adjoint to the same relative Ext functors. More precisely, writing S^i for either side and V for any representation of K,

$$(6.32) \qquad \mathrm{Hom}_K(V,S^i) = \mathrm{Ext}^i_{(q\cap\mathfrak{k},L\cap K)}(V,W).$$

When S^i comes from the right side in the proposition, this is a generalization of Proposition 1.27. (It goes back to [Bott, 1957].) For S^i coming from the left side in the proposition, it follows from Proposition 6.25 and a version of Shapiro's lemma.

Definition 6.33 (cf. [Vogan, 1981], Theorem 6.3.12). In the setting of Definition 6.20, the *order of vanishing along $\mathfrak{k}$ filtration* is the decreasing filtration of X (by ($\mathfrak{k}$,L∩K)-submodules)

$$(6.33)(a) \qquad X_n = \{\phi \in X|\ \phi(u) = 0, \ \text{all } u \in U_n(\mathfrak{g})U(\mathfrak{k})\}.$$

In particular, X_0 is the subspace of maps vanishing on $U(\mathfrak{k})$; so

$$(6.33)(b) \qquad X/X_0 \cong \mathrm{Hom}_{q\cap\mathfrak{k}}(U(\mathfrak{k}),W)_{L\cap K}$$
$$= \mathrm{pro}^{(\mathfrak{k},L\cap K)}_{(q\cap\mathfrak{k},L\cap K)}(W).$$

More generally, it is easy to check that

(6.33)(c) $X_{n-1}/X_n \cong \mathrm{Hom}_{q \cap \mathfrak{k}}(U(\mathfrak{k}), S^n((\mathfrak{s}/q \cap \mathfrak{s})\,)\, \otimes\, W)_{L \cap K}$

$\qquad\quad = \mathrm{pro}^{(\mathfrak{k}, L \cap K)}_{(q \cap \mathfrak{k}, L \cap K)}(S^n(\mathfrak{s}/q \cap \mathfrak{s})^* \otimes W).$

The idea now is to use the order of vanishing along $\mathfrak{k}$ filtration to compute $\Gamma^i X$ as a representation of K. Because of Lemma 6.23, this is permitted in principle. Because the filtration is infinite, there is a minor technical problem about convergence of a spectral sequence. It can be handled by a simple trick, however. The result is

THEOREM 6.34 (generalized Blattner formula); (see [Vogan, 1981], Theorem 6.3.12). *In the setting of Theorem 6.8 and Definition 6.19, assume that* Z *has finite length. Write* $\mathcal{W}_n$ *for the holomorphic vector bundle on* $K/L \cap K$ *corresponding to the* $(q \cap \mathfrak{k}, L \cap K)$*-module*

$$S^n((\mathfrak{s}/q \cap \mathfrak{s})^*)\, \otimes\, W.$$

a) $\mathcal{R}^j(Z) = 0$ *for* j *greater than* S.

b) *All the* $\mathcal{R}^j(Z)$ *are admissible (Definition 2.11).*

c) *As virtual representations of* K,

$$\sum_j (-1)^j \mathcal{R}^j(Z) = \sum_{j,n} (-1)^j H^j(K/L \cap K, \mathcal{W}_n).$$

The right side of the formula in the theorem may be computed
using Bott's generalization of the Borel-Weil theorem (cf.
[Bott, 1957)].

The corner of the spectral sequence in the proof of
this theorem arises as follows. The order of vanishing filtration gives a surjective map

$$\text{pro}_{(q,L\cap K)}^{q,K}(W) \to \text{pro}_{(q\cap \ell,L\cap K)}^{\ell,K}(W).$$

Now apply the derived functors of Γ (using Lemma 6.23).
We get

COROLLARY 6.35. *In the setting of Definitions 6.20 and*
6.30, there is a surjective map of K-modules

$$\mathscr{R}^S(Z) \to (\mathscr{R}_{q\cap \ell, L\cap K})^S(Z).$$

We return now to our discussion of the proof of Theorem
6.8. Part (b) follows from Theorem 6.34(b). For (d)
through (g), we may as well assume (perhaps after replacing
Z by $Z \oplus Z^h$) that Z carries a non-degenerate invariant
Hermitian form. By Proposition 6.29, $X = \text{pro}(W)$ does as
well. Notice that S is half the dimension of $K/L\cap K$. The
decomposition

$$\ell/\mathfrak{l} \cap \ell \cong (u \cap \ell) \oplus (u^- \cap \ell),$$

and the invariant pairing between the two summands induced
by our fixed bilinear form on $\mathfrak{g}$, show that $L \cap K$ acts

trivially on the top exterior power of $\mathfrak{k}/\mathfrak{l} \cap \mathfrak{k}$. By Theorem 6.28, $\mathscr{R}^S(Z)$ carries a non-degenerate form, as required by Theorem 6.8(e). In addition,

$$\mathscr{R}^j(Z)^h = \mathscr{R}^{2S-j}(Z).$$

If j is greater than S, the left side is zero by Theorem 6.34(a). If j is less than S, the right side is zero for the same reason. Theorem 6.8(d) follows. The statement about definiteness in Theorem 6.8(e) is rather hard. Proofs may be found in [Vogan, 1984] or [Wallach, 1984].

For the remaining statements, notice that Theorem 6.8(d) and Theorem 6.34(c) compute $\mathscr{R}^S(Z)$ as a representation of K (under the hypothesis of Theorem 6.8(d)). The non-vanishing statement in Theorem 6.8(f) is proved using Corollary 6.35, by exhibiting a representation of L ∩ K that is not killed by $(\mathscr{R}_{\mathfrak{q}\cap\mathfrak{k},L\cap K})^S$. For (g), [Schmid, 1975] contains a characterization of any discrete series representation in terms of its restriction to K. Theorem 6.34 allows one to check that $\mathscr{R}^S(\tau)$ satisfies the conditions of that characterization.

To conclude this chapter, we discuss the unitary representation theory of $GL(n,\mathbb{R})$. Details may be found in [Vogan, 1986b]. To see how the theory differs from the case of $GL(n,\mathbb{C})$, we begin with a series of representations studied by B. Speh in [Speh, 1981]. Suppose $n = 2m$, and consider the Levi subgroup

(6.36)(a) $L = GL(m,\mathbb{C})$

of G (cf. (5.3)). Recall that X denotes the element of $\mathfrak{g}_0$ corresponding to multiplication by i (when $\mathbb{R}^{2m}$ is identified with $\mathbb{C}^m$). The eigenvalues of ad(X) turn out to be 0 and $\pm 2i$. Put

(6.36)(b) $\mathfrak{u} = +2i$ eigenspaces of ad(X)

(6.36)(c) $\mathfrak{q} = \mathfrak{l} + \mathfrak{u}$.

Then $\mathfrak{q}$ is a parabolic subalgebra of $\mathfrak{g}$, opposite to its complex conjugate. If we take the Cartan involution θ on G to be inverse transpose, then θX is X; so $\mathfrak{q}$ is θ-stable.

For each integer k, define a unitary character χ_k of L by

(6.37) $\chi_k(\ell) = \det(\ell)/|\det(\ell)|$.

(Here we refer to the determinant function on $GL(m,\mathbb{C})$.) In the notation of Definition 5.7, one can calculate that

(6.38)(a) $2\rho(\mathfrak{u}) = \chi_{2m}$

This has a square root (namely χ_m). It follows that the metaplectic cover of $\tilde{L}$ of L (Definition 5.7) is isomorphic to $L \times \mathbb{Z}/2\mathbb{Z}$, and that metaplectic representations of $\tilde{L}$ may be identified with representations of L. (To use

this identification in Definition 6.20, for example, one sim-
ply replaces tensoring with $\mathbb{C}_{\rho(u)}$ by tensoring with χ_m.)

We are going to consider the representations

(6.39) $$X_k = \mathscr{R}^S(\chi_k)$$

of $GL(2m,\mathbb{R})$. Theorem 6.34 suggests looking first at K,
which in this case is the orthogonal group $O(2m)$. Because
of Proposition 6.31, the following result is a special case
of the Bott-Borel-Weil theorem (cf. [Bott, 1957]).

LEMMA 6.40. *In the setting above, put*

$$V = (\mathscr{R}_{q\cap\mathfrak{k},L\cap K})(\chi_{k-1}).$$

a) If $k < 0$, then V_k is zero.

*b) If $k = 0$, then V_k is the sum of the trivial and
the determinant characters of $O(2m)$.*

*c) If $k = 1$, then V_k is the representation of
$O(2m)$ on $\Lambda^m(\mathbb{C}^{2m})$.*

*d) If $k \geq 1$, then V_k is an irreducible representa-
tion of $O(2m)$. It is the kth Cartan power of V_1. In
appropriate standard coordinates, it has highest weight
$(k,\ldots,k)$.*

The shift by -1 in the definition of V_k is made to simplify the remaining statements.

Corollary 6.35 now guarantees that X_k is non-zero as long as k is at least -1. On the other hand, the condition in Theorem 6.8(d) amounts to

$$k \geq m - 1$$

in the present case. There is therefore a range of values of k for which the general theory does not guarantee nice behavior of X_k, but for which the restriction to K looks reasonable. It is possible to improve substantially on the general theory, to make it cover most of this range. We will return to this point in Chapter 13 (Theorem 13.6). For now, we simply state what happens here.

PROPOSITION 6.41. *In the setting of* (6.36) *and* (6.37), *the Harish-Chandra module*

$$X_k = I_{q,L}(x_k)$$

for $GL(2m,\mathbb{C})$ *(notation* (6.9)) *is irreducible and infinitesimally unitary for* $k \geq 0$. *It contains the representation* V_{k+1} *of* $O(2m)$ *(cf. Lemma* 6.40) *as its lowest K-type in the sense of* [Vogan, 1981].

When k *is* -1, X_k *is neither irreducible nor infinitesimally unitary.*

The main point here is that X_k is unitary for non-negative k. This is due to Speh. (One must bear in mind that this result was proved before Theorem 6.8(e) was available.) What Speh did was show that X_k actually appears as a constituent of the (unitary) representation of $GL(2m, \mathbb{R})$ on a certain space of square-integrable automorphic forms.

When k is zero, X_k is equivalent to a representation induced from the real parabolic subgroup of $GL(2m, \mathbb{R})$ with Levi factor $GL(m, \mathbb{R}) \times GL(m, \mathbb{R})$, by a certain one-dimensional unitary character. For k positive, X_k is not equivalent to any representation induced from a proper real parabolic subgroup of G. This behavior shows the differences in the unitary representation theory of $GL(n, \mathbb{R})$ and $GL(n, \mathbb{C})$. It is necessary to introduce cohomological induction to get all the representations; and when that is done, the problem of finding all equivalences among the various constructions becomes rather delicate. We will not try to formulate the answer precisely here, but the following theorem contains the main point.

THEOREM 6.42 ([Vogan, 1986b]). *Any irreducible unitary representation of* $GL(n, \mathbb{R})$ *may be obtained by iterating the processes of parabolic induction (Definition 4.1) and coho-*

mological parabolic induction (Definition 6.20), starting

from two kinds of representations: one-dimensional unitary

characters, and Stein complementary series (cf. Theorem

4.23).

We actually need the Stein complementary series over both $\mathbb{R}$

and $\mathbb{C}$. As was pointed out after Theorem 4.23, the choice

of ground field hardly affects their construction.

Interlude

THE IDEA OF UNIPOTENT REPRESENTATIONS

Theorem 6.42 is a model of what one would like to know for any reductive group: that any unitary representation is obtained by systematic processes from a small number of building blocks. The systematic processes should certainly include unitary induction from real parabolic subgroups (Definitions 3.8 and 4.1), and cohomological parabolic induction (Definition 6.20). For our purposes, we will regard the formation of complementary series (as in Theorem 4.23) as another "systematic process," even though the limits of its applicability are not nearly so well understood as in the first two cases. We are therefore led to

Problem I.1. For each reductive group G, describe a nice class $\mathscr{B}(G)$ of unitary representations of G, with the following property. Let π be any irreducible unitary repre-

sentation of G. Then there is a Levi subgroup L of G
(Definition 5.1), and a unitary representation π_L in
$\mathscr{B}(L)$, such that π is obtained from π_L by a complementary
series construction, followed by real parabolic induction,
followed by cohomological parabolic induction.

(The letter $\mathscr{B}$ may be taken to stand for building block.)
In the sense intended here, the Stein complementary series
for $GL(2m,\mathbb{C})$ are obtained from the one-dimensional unitary
character $\xi(k,k)\otimes(it,it)$ of $GL(m)\times GL(m)$ (cf. Theorem 4.23).
(Replacing this by the non-unitary character $\xi(k,k)$ $\otimes$
$(\sigma+it,-\sigma+it)$ (from which one actually induces in the end)
is regarded as part of the complementary series construction.)

Dan Barbasch has pointed out that this problem is
almost certainly phrased too optimistically in at least one
important respect. Under special conditions, a non-unitary
Hermitian representation can give rise to unitary representations by a complementary series kind of procedure: one
deforms an indefinite induced inner product through some
poles until it becomes definite. If this is really the most
natural construction of these representations, then they
will not fit into the scheme proposed in Problem I.1. It is

not yet clear how to deal with this difficulty. It should
not interfere with the less ambitious and more precise
conjectures stated later, however.

For G a product of copies of $GL(n,\mathbb{R})$ or $GL(n,\mathbb{C})$,
Theorems 4.28 and 6.42 suggest that one can take $\mathcal{B}(G)$ to
consist of the set of one-dimensional unitary characters of
G. At any rate we must include these char-acters in $\mathcal{B}(G)$;
for they cannot be obtained from any smaller group in the
manner suggested.

This immediately suggests the hope that one might take
$\mathcal{B}(G)$ to consist of the unitary characters of G in general. This hope fails for the first time when G is the
group $Sp(2n,\mathbb{C})$, of linear transformations of $\mathbb{C}^{2n}$ preserving a non-degenerate symplectic form. G has a beautiful
unitary representation ω on $L^2(\mathbb{C}^n)$, called the *metaplectic representation*. It was originally constructed in
[Shale, 1962]; a nice account of it may be found in [Howe,
1980]. (The analogous representation for the real field is
discussed briefly in Example 11.26.) The representation ω
is a direct sum of two irreducible pieces ω^+ and ω^-. If
n is at least two, neither ω^+ nor ω^- may be constructed
by real or cohomological parabolic induction from a proper
parabolic subgroup; and neither piece is any kind of complementary series. (If n is 2, ω^+ appears as a subquotient

at the end of a complementary series. Such a "construction"
proves that ω^+ is unitary but does not compute its character; Problem I.1 is not intended to allow it.) We are therefore obliged (by the conditions of Problem I.1) to allow $\omega^{\pm}$
in $\mathscr{B}(\mathrm{Sp}(2n,\mathbb{C}))$.

Having admitted $\omega^{\pm}$ to $\mathscr{B}(\mathrm{Sp}(2n,\mathbb{C}))$ for n at least
2, we would be hard pressed to construct a nice theory ex-
cluding it for n equal to 1. But a symplectic form on $\mathbb{C}^2$
is just a volume form; so $\mathrm{Sp}(2,\mathbb{C})$ is just $\mathrm{SL}(2,\mathbb{C})$. This
suggests that even for $\mathrm{SL}(n)$, we ought to allow $\mathscr{B}(G)$ to
contain more than the unitary characters. To put it another
way, it seems unreasonable to choose $\mathscr{B}(G)$ to be the smallest set solving Problem I.1. Rather, we will seek some
interesting larger set, defined more or less independently
of Problem I.1, with the hope that it will solve that problem by magic. A little more precisely, we pose

   Problem I.2. For each semisimple Lie group G, describe a finite set $\mathscr{U}(G)$ of irreducible unitary representations, with the following properties.

   i) Suppose π is a representation of G. Write π_0
for the restriction of π to the identity component G_0 of
G. Then π belongs to $\mathscr{U}(G)$ if and only if each constituent of π_0 belongs to $\mathscr{U}(G_0)$.

ii) Suppose G is reductive. Define $\mathscr{B}(G)$ to consist of all irreducible unitary representations of G whose
restriction to the commutator subgroup G' of G is a sum
of elements of $\mathscr{U}(G')$. Then $\mathscr{B}(G)$ should satisfy the
condition in Problem I.1.

iii) Suppose G is semisimple. Then any representation of G which is trivial on the identity component
belongs to $\mathscr{U}(G)$.
Suppose G is semisimple, and $\mathscr{U}(G)$ is defined. A unitary
representation of G will be called *unipotent* if it is a
direct sum of elements of $\mathscr{U}(G)$.

The requirements posed in Problem I.2 are much too weak
to determine what representations ought to be considered unipotent. In the course of the next five chapters, we will
consult a series of oracles, with the aim of adding to the
list of requirements. Finally we would like it to be so
long – even without condition (I.2)(ii) – as to specify
$\mathscr{U}(G)$ completely. (This goal will not be attained.) The
question of whether (I.2)(ii) holds is then a separate
issue, essentially equivalent to the classification of the
unitary representations of all reductive groups.

Chapter 7

FINITE GROUPS AND UNIPOTENT REPRESENTATIONS

The first oracle that we will consult is Lusztig's work
[Lusztig, 1984] on (complex) representations of finite
Chevalley groups. In the first half of this chapter (through
Corollary 7.16) G will always denote a reductive algebraic
group defined and split over the finite field F_q (with q
elements). We will require G to have connected center.
The group $G(F_q)$ is a finite Chevalley group. Write

(7.1) $B(e) = T(e)N(e)$

for a Levi decomposition defined over F_q of a Borel subgroup of G. (The "e" represents the identity element of
the Weyl group; the notation will be generalized in a
moment.) Because G is assumed to be split, $T(e)(F_q)$ is
a product of copies of the multiplicative group of F_q.

Two examples will illustrate most of the main points:
the groups

(7.2)(a) $G = GL(n)$,

of all invertible $n \times n$ matrices; and

(7.2)(b) $G = SO(2n+1)$,

consisting of $(2n+1) \times (2n+1)$ matrices preserving the (maximally isotropic) quadratic form

$$Q(x_0, x_1, \ldots, x_{2n}) = (x_0)^2 + x_1 x_2 + x_3 x_4 + \ldots + x_{2n-1} x_{2m}$$

on $(2n+1)$-dimensional space. In the first case, we can take $T(e)$ to consist of all diagonal matrices and $N(e)$ to consist of strictly upper triangular matrices. In the second, $T(e)$ may be taken to be diagonal matrices of the form

$$d(1, z_1, (z_1)^{-1}, \ldots, z_n, (z_n)^{-1});$$

we are using the notational convention of (4.19). We omit the description of $N(e)$.

Just as for real groups, representations are associated roughly to characters of Cartan subgroups. More precisely, fix a maximal torus T of G, defined over F_q, and a character θ of $T(F_q)$. In [Deligne-Lusztig, 1976], a virtual representation

(7.3) $R_T(\theta)$

of $G(F_q)$ is defined. This construction is entirely analogous to the various kinds of parabolic induction introduced earlier for real groups. For example, if $T = T(e)$, then

(7.4) $R_T(\theta) = \text{Ind}_{B(e)(F_q)}^{G(F_q)}(\theta).$

The representations $R_T(\theta)$ are irreducible for "most"
choices of θ; and every irreducible representation of $G(F_q)$
occurs in one of them. To describe all the irreducible
representations of $G(F_q)$, it therefore suffices to decompose all of the $R_T(\theta)$ into irreducible representations.
The most difficult case, and the one to which Lusztig eventually reduces the general case, is when θ is trivial.

Definition 7.5. A *unipotent representation* of the finite
Chevalley group $G(F_q)$ is one occurring in some $R_T(1)$.

Because the Frobenius automorphism acts trivially on
the rational characters of the split torus $T(e)$ – that is
what split means – it must act by a G-inner automorphism on
the characters of any torus defined over F_q. That is, it
must act by some element of the Weyl group W for G. This
turns out to define a bijection between $G(F_q)$-conjugacy
classes of maximal tori defined over F_q and conjugacy
classes in W. Write $T(w)$ for a torus in the class corresponding to an element w of W; this notation is consistent with the earlier choice of $T(e)$. Put

(7.6) $R_w = R_{T(w)}(1)$.

By the preceding remarks, R_w depends only on the conjugacy
class of w.

For w equal to the identity element e, (7.4) shows that R_e is equal to the representation of $G(F_q)$ on the space of functions on $G(F_q)/B(e)(F_q)$. A fairly straightforward analysis of this space leads to the following classical result.

PROPOSITION 7.7. *The irreducible representations of* $G(F_q)$ *occurring in* R_e *are in a natural one-to-one correspondence with the irreducible representations of the Weyl group of* G.

The Weyl group arises because it parametrizes the double cosets of $B(e)(F_q)$ in $G(F_q)$. Write

(7.8)(a) $$X(\sigma) \in G(F_q)^{\wedge}$$

for the irreducible representation corresponding to a representation σ in $\hat{W}$. It is convenient to twist the most obvious version of this parametrization by tensoring with the sign representation of W. With this convention,

(7.8)(b) $X(\text{sgn}) =$ trivial representation of $G(F_q)$

(7.8)(c) $X(1) =$ Steinberg representation of $G(F_q)$.

We now have a family of virtual characters of $G(F_q)$ parametrized by conjugacy classes in W, and a family of irreducible representations parametrized by $\hat{W}$. It is natural to hope that these might be related by the character table of W. Explicitly, define for σ in $\hat{W}$

(7.9)(a) $R_\sigma = |W|^{-1} \sum_{w \in W} \text{tr } \sigma(w) R_w.$

Notice that R_σ is only a rational (as opposed to integral) combination of irreducible characters. In case G is $GL(n)$, R_σ is simply equal to $X(\sigma)$. In general, the situation is more complicated; $X(\sigma)$ occurs in R_σ, but other representations do as well. The R_σ are have far fewer irreducible constituents than the R_w, but they are not irreducible or disjoint. Nevertheless, we always have

(7.9)(b) R_1 = Steinberg representation of $G(F_q)$

R_{sgn} = trivial representation of $G(F_q)$.

Definition 7.10. Let LR-*equivalence* be the equivalence relation on $\hat{W}$ generated by the requirement that $\sigma \sim_{LR} \tau$ if R_σ and R_τ share an irreducible constituent. An equivalence class is called a *double cell*.

A unipotent representation X of $G(F_q)$ (Definition 7.5) is said to be *attached* to the double cell $\mathscr{C}$ if X occurs in R_σ, for some representation σ of W belonging to $\mathscr{C}$. Set

$\mathcal{U}(\mathcal{C})$ = set of irreducible representations of $G(F_q)$

  attached to $\mathcal{C}$

  $\supset \{X(\sigma) \mid \sigma \in \mathcal{C}\}$.

(This last containment requires proof.)

It is shown in [Lusztig, 1984] that this definition of double cells is equivalent to the one implicit in [Kazhdan-Lusztig, 1979], which is made in terms of Hecke algebras and the Kazhdan-Lusztig polynomials. This makes possible some other formulations of Definition 7.10. None is particularly simple or geometric, however.

Two particular double cells are worthy of special mention. First, the trivial representation is a double cell by itself, and the only unipotent representation attached to it is the Steinberg representation. Similarly, the sign representation of W is a double cell, attached only to the trivial representation of G.

Definition 7.10 partitions the unipotent representations according to double cells in W. The most important remaining step is to understand the set of double cells; but we intend to postpone that question and consider instead the fine structure of the sets $\mathcal{U}(\mathcal{C})$.

Definition 7.11 (cf. [Lusztig, 1984], section 4.3). Suppose
A is a finite group. Consider the set of all pairs $(x, \mathscr{C})$,
with x in A and $\mathscr{C}$ an irreducible representation of the
centralizer $Z_A(x)$ of x in A. The group A acts on the
set of such pairs, as follows. If g is in A, then compos-
ing $\mathscr{C}$ with conjugation by g defines a representation π^g
of

$$g(Z_A(x))g^{-1} = Z_A(gxg^{-1}).$$

Define $\mathscr{M}(A)$ to be the set of A orbits of pairs (x, π)
as above. Notice that $\mathscr{M}(A)$ contains both $\hat{A}$ (as the set
of pairs (e, π); the action of A on such pairs is trivial)
and the set of conjugacy classes in A (as the set of orbits of pairs of the form $(x, 1)$). If A is abelian, then

$$\mathscr{M}(A) = A \times \hat{A}.$$

THEOREM 7.12 ([Lusztig, 1984], Chapter 4). *Let G be a
connected reductive algebraic group, defined and split over
the finite field F_q. Assume that G has connected center.
Write W for the Weyl group of G. Fix a double cell $\mathscr{C}$
of representations of W (Definition 7.10). Then one can
attach to $\mathscr{C}$ a finite group $A(\mathscr{C})$ with the following property: the set $\mathscr{U}(\mathscr{C})$ of unipotent representations of $G(F_q)$
attached to $\mathscr{C}$ (Definition 7.10) is in one-to-one correspondence with $\mathscr{M}(A(\mathscr{C}))$ (Definition 7.11).*

This is a very deep result, involving both powerful general methods and extensive case-by-case calculations. The group $A(\mathscr{C})$ is defined case by case, although Lusztig gives a general definition *a posteriori* in Chapter 13 of [Lusztig, 1984].

In the correspondence given by Theorem 7.12, write

$$(7.13)(a) \qquad X(x,\pi) \qquad (x \in A(\mathscr{C}), \; \pi \in (Z_{A(\mathscr{C})}(x))^{\wedge}$$

for the representation of $G(F_q)$ corresponding to the equivalence class of the pair (x,π). We have already assigned to each σ in $\mathscr{C}$ a unique representation $X(\sigma)$ in $\mathscr{U}(\mathscr{C})$ (notation (7.8)). This provides an injective map

$$(7.13)(b) \qquad\qquad \mathscr{C} \to \mathscr{M}(A(\mathscr{C})),$$

which we write as

$$(7.13)(c) \qquad\qquad \sigma \to m(\sigma).$$

It turns out that every conjugacy class in $A(\mathscr{C})$ (regarded as an element of $\mathscr{M}(A(\mathscr{C}))$; see Definition 7.11) is of the form $m(\sigma)$. If x is in $A(\mathscr{C})$, we can therefore define

$$(7.13)(d) \qquad\qquad \sigma(x) \in \hat{W}.$$

Lusztig computes the multiplicity of each irreducible representation in each $R_{\sigma(x)}$ ([Lusztig, 1984], Theorem 4.23). A special case is

THEOREM 7.14. *In the setting of Theorem 7.12, fix* $x \in$
$A(\mathscr{C})$, *and an irreducible representation* π *of* $A(\mathscr{C})$. *Then
the irreducible representation* $X(e,\pi)$ *of* $G(F_q)$ *(Theorem
7.12) occurs in* $R_{\sigma(x)}$ *(cf. (7.13)(d) and (7.9)) with multiplicity*

$$|Z_{A(\mathscr{C})}(x)|^{-1} \mathrm{tr}(\pi(x)).$$

The formula here suggests using Fourier inversion on the
finite group $A(\mathscr{C})$. Specifically, write $[x]$ for the conjugacy class of an element x of $A(\mathscr{C})$, and $[A(\mathscr{C})]$ for the
set of all conjugacy classes. Fix an irreducible representation δ of $A(\mathscr{C})$, and define

(7.15) $Z_\delta = \displaystyle\sum_{[x]\in[A(\mathscr{C})]} \mathrm{tr}(\delta^*(X))R_{\sigma(x)}$

(notation (7.13)(d) and (7.9); δ^* denotes the representation dual to δ). In terms of the Z_δ, Theorem 7.14 may be
reformulated as

COROLLARY 7.16. *In the setting of Theorem 7.12, fix an
irreducible representation* δ *of* $A(\mathscr{C})$, *and define a rational virtual character* Z_δ *(of* $G(F_q)$*) by (7.15). Then
for any irreducible representation* π *of* $A(\mathscr{C})$, *the representation* $X(e,\pi)$ *of* $G(F_q)$ *occurs in* Z_δ *with multipli-*

city one if π *is equivalent to* δ*, and multiplicity zero otherwise.*

Corollary 7.16 does not say that the Z_δ are irreducible characters; they will in fact contain many unipotent representations. (Except for a sign appearing only in E_7 and E_8, the multiplicity of $X(x,\pi)$ in Z_δ turns out to be the multiplicity of δ in the representation

$$\text{Ind}_{Z(x)}^{A(\mathscr{C})}(\pi)$$

of $A(\mathscr{C})$.) The problem is that there are more unipotent representations than R_w's; so one cannot hope to find formulas for the irreducible representations in terms of the R_w's.

These results on finite Chevalley groups suggest that any representation of a real reductive group obtained (by a combination of real and cohomological parabolic induction) from a trivial character of a Cartan subgroup ought to be regarded as unipotent. We will certainly require this; but the resulting set of representations is much less rich than the analogy with finite groups would suggest.

To explain this in more detail, let us consider the case of complex groups. For the rest of this chapter, we will therefore fix

(7.17)(a)   G = a complex connected reductive

algebraic group

In addition to the usual structure we have introduced for general reductive groups (such as K, θ, and so on), we choose

(7.17)(b)   B = TAN a Langlands decomposition

of a Borel subgroup.

It follows automatically that T is a maximal torus in K. Write

(7.17)(c)   H = TA.

Then H is a representative of the unique conjugacy class of Cartan subgroups of G. One constructs representations of G from characters of H by induction from B (Definition 4.1). Finally, set

(7.17)(d)   W = W(G,H),

the Weyl group of H in G (cf. Definition 5.1).

Because of the presence of the modular function δ in the definition of induction, there are two natural candidates for the "right" generalization to complex groups of the representation R_e (cf. (7.6)) for a finite Chevalley group. The first is

(7.18)(a) $I(0) = \text{Ind}_B^G(1)$.

This is a unitary representation of G, belonging to the minimal principal series (Example 3.20). Some of the analysis leading to Proposition 7.7 may be applied to I(0).

This was begun in [Bruhat, 1956]. One finds a family of intertwining operators

(7.18)(b) $A(w)$: $I(0) \to I(0)$,

parametrized by the Weyl group W. (Up to a normalizing constant, $A(w)$ is just the operator $A(w:0)$ of Theorem 4.11.) It can be shown that the map sending w to $A(w)$ is a representation of W, and that every intertwining operator for $I(0)$ is a linear combination of the various $A(w)$. (Bruhat's original results were not this complete, but they were of this nature.) It follows easily that the irreducible constituents of $I(0)$ are parametrized by certain representations of W: those that appear in the representation (7.18)(b) of W.

So far the analogy with Proposition 7.6 appears perfect. Unfortunately, it turns out that the operators $A(w)$ are all the identity ([Kostant, 1969]). (This is by no means a trivial fact; the definition of $A(w)$, which was omitted, is quite subtle.) Therefore $I(0)$ is irreducible; the beautiful theory for finite Chevalley groups that begins with Proposition 7.6 has no counterpart here. We can call $I(0)$ unipotent; but (since $I(0)$ is not the trivial representation of G unless G is abelian) condition (I.2)(iii) (in the preliminary requirements for unipotent representations formulated in the Interlude) will not allow it to be the *only* unipotent representation.

The second possible analogue of R_e is the space of functions on G/B. By Definition 4.1, this is

$$(7.19) \qquad I(-\rho) = \text{Ind}_B^G(\delta^{-\frac{1}{2}}).$$

The difficulty with this representation is that it is not naturally unitary (since G/B admits no G-invariant measure); and in fact $I(-\rho)$ generally has some non-unitary composition factors. On the other hand, $I(-\rho)$ *does* contain the trivial representation (on the subspace of constant functions). It is therefore reasonable to try to decompose $I(-\rho)$ into irreducibles, looking for something like Proposition 7.7. To explain the situation, we need a slightly more detailed description of the characters of H.

Recall that H, as a Cartan subgroup of a complex group, is itself complex; so its Lie algebra $\mathfrak{h}_0$ is also complex. On the other hand, we are interested in the underlying real Lie algebra, and its complexification $\mathfrak{h}$.

LEMMA 7.20. *In the setting (7.17), there is a decomposition*

a) $\mathfrak{h} = \mathfrak{h}^L + \mathfrak{h}^R$; *here $\mathfrak{h}^L$ and $\mathfrak{h}^R$ are each naturally isomorphic to $\mathfrak{h}_0$. In this decomposition, $\mathfrak{t}$ corresponds to the skew diagonal, and $\mathfrak{a}$ to the diagonal:*

b) $\mathfrak{t} = \{(X,-X) \mid X \in (\mathfrak{h}_0)^*\}.$

c) $\alpha = \{(X,X) \mid X \in (\mathfrak{h}_0)^*\}.$

These definitions provide identifications of $\mathfrak{t}$ *and* α
with $\mathfrak{h}_0$. *Fix* λ *and* μ *in* $(\mathfrak{h}_0)^*$. *Then the weight*

$$(\lambda,\mu) \in \mathfrak{h}^*$$

exponentiates to H *if and only if* $\lambda-\mu$ *exponentiates to*
T. *(Here we use the identification above to regard* $\lambda-\mu$ *as*
an element of $\mathfrak{t}^*$.*)*

Finally, *the Weyl group* $W_{\mathbb{C}}$ *of* $\mathfrak{h}$ *in* $\mathfrak{g}$ *(Definition*
5.1) may be identified with a product of two copies of W,
acting separately on the two factors in (a) *above:*

d) $W_{\mathbb{C}} = W^L \times W^R.$

The real Weyl group W(G,H) *is the diagonal subgroup.*

This is very easy. A more complete description of the various decompositions may be found in [Vogan, 1981], section
7.1.

Definition 7.21. In the setting (7.17), fix λ and μ in
$(\mathfrak{h}_0)^*$. Assume that $\lambda-\mu$ exponentiates to T, so that we
get a character $\mathbb{C}_{(\lambda,\mu)}$ of H (cf. Lemma 7.20). Define

$$I(\lambda,\mu) = \mathrm{Ind}_B^G(\mathbb{C}_{\lambda,\mu})$$

(cf. (4.10)), a principal series representation of G. By
(4.2),

$$I(\lambda,\mu)\big|_K = \mathrm{Ind}_T^K(\mathbb{C}_{\lambda-\mu}).$$

By Theorem 1.30, this restriction to K contains the representation $F_{\lambda-\mu}$ of extremal weight $\lambda-\mu$ exactly once. Write

$$J(\lambda,\mu)$$

for the unique irreducible subquotient of $I(\lambda,\mu)$ that contains the representation $F_{\lambda-\mu}$ of K.

        Write

(7.22)(a) $\Delta_0 = \Delta(\mathfrak{g}_0,\mathfrak{h}_0)$

for the set of roots of $\mathfrak{h}_0$ in $\mathfrak{g}_0$ (regarded as complex

Lie algebras). In the identification of Lemma 7.20(a), the

roots of $\mathfrak{h}$ in $\mathfrak{g}$ are

(7.22)(b) $\Delta = \Delta^L \cup \Delta^R.$

Write

(7.22)(c) $\Delta^+ = (\Delta^L)^+ \cup (\Delta^R)^+$

for the set of positive roots defined by n. Finally, write

(7.22)(d) $\rho_{\mathbb{C}} = (\rho,\rho) \in \mathfrak{h}^*$

for half the sum of the positive roots; here

(7.22)(e) $\rho \in (\mathfrak{h}_0)^*$

is half the sum of the roots for $\mathfrak{h}_0$ in $\mathfrak{g}_0$.

        Using the notation of Definition 7.21 and (7.22), we

can write the representation $I(-\rho)$ of (7.19) as

(7.23) $I(-\rho) = I(-\rho,-\rho).$

Recall that we are seeking a formula analogous to that of

7.9(a) for the trivial representation. Here it is.

PROPOSITION 7.24 (Zuckerman; see [Vogan, 1981], Proposition

2.2.10). *Suppose* G *is a complex connected reductive alge-*

braic group. *In the notation of Definition 7.21, there is*

an equality of virtual representations of G

$$\mathbb{C} = \sum_{w \in W} \text{sgn}(w) \ I(-\rho, -w\rho).$$

The proof is largely formal; the result is essentially equi-

valent to the Weyl character formula (Theorem 1.40). It is

necessary only to understand some kind of character theory

for (g,K)-modules.

To understand this and subsequent formulas properly, we

need some further information about the representations

$I(\lambda,\mu)$.

LEMMA 7.25 (Zhelobenko; see [Duflo, 1975]). *In the setting*

of Definition 7.21, two representations $I(\lambda,\mu)$ *and*

$I(\lambda',\mu')$ *have the same irreducible composition factors -*

that is, the same image in the Grothendieck group of finite

length (g,K)-modules - *if and only if there is a* w *in* W

such that

$$w\lambda = \lambda' \qquad \text{and} \qquad w\mu = \mu'.$$

Modulo these equivalences, the $I(\lambda,\mu)$ *form a basis of the Grothendieck group.*

Example 7.26. Suppose G is $SL(2,\mathbb{C})$. Then it turns out that

$$I(-\rho,\rho) = J(-\rho,\rho)$$

is irreducible. By Lemma 7.20, the character from which it is induced is

$$-\rho - \rho = -2\rho$$

on the compact part T of the Cartan subgroup; it is

$$-\rho + \rho = 0$$

on the split part A. In particular, it is unitary; so $I(-\rho,\rho)$ is a unitary principal series representation. Its restriction to K = SU(2) consists of every odd-dimensional representation except the trivial one.

The representation $I(-\rho,-\rho)$ (on functions on the Riemann sphere G/B – see (7.23)) is induced from a charac-ter trivial on T, but not trivial on A. Its restriction to K consists of all odd-dimensional representations of SU(2). Because the constant functions transform by the trivial representation of K, Definition 7.21 gives

$$J(-\rho,-\rho) = \mathbb{C}.$$

The quotient representation (on functions modulo constants)

turns out to be irreducible and isomorphic to $J(-\rho,\rho)$. In the Grothendieck group, therefore,

$$I(-\rho,-\rho) = J(-\rho,-\rho) + J(-\rho,\rho).$$

Consequently

$$\mathbb{C} = J(-\rho,-\rho) = I(-\rho,-\rho) - I(-\rho,\rho),$$

which is the formula of Proposition 7.24 in this case.

Proposition 6.7 and Lemma 7.25 may be combined to give the following result.

PROPOSITION 7.27. *In the setting of Definition 7.21, fix* λ *and* μ *so that* $\lambda-\mu$ *exponentiates to* T. *Define*

$$W(\mu) = \{w \in W \mid w\mu - \mu \text{ is a sum of roots}\}$$

$$W_\mu = \text{stabilizer of } \mu \text{ in } W.$$

Regarded as an element of the Grothendieck group of finite length Harish-Chandra modules, $I(\lambda,w\mu)$ *depends only on the double coset*

$$W_\lambda w W_\mu.$$

Let X *be any irreducible Harish-Chandra module for* G, *on which* $Z(G)$ *and* $\mathcal{Z}(\mathfrak{g})$ *(Definition 6.3) act as in* $I(\lambda,\mu)$. *Then there is a unique expression*

$$X = \sum_{w \in W_\lambda \backslash W(\mu) / W_\mu} a(w)I(\lambda,w\mu),$$

*with a(w) an integer; this formula is to be interpreted in
the Grothendieck group.*

This proposition allows us to regard the various
$I(\lambda, w\mu)$ as roughly analogous to the R_w of (7.6), and
Proposition 7.24 is an encouraging analogue of (7.9)(b).
The analogy with (7.9)(a) would suggest considering such
virtual representations as

$$\sum I(-\rho, -\rho).$$

This turns out to be a rather bad idea, however. In Example
7.26, it gives

$$2J(-\rho, \rho) + J(-\rho, -\rho),$$

which is not a simpler object than the individual principal
series representations. Part of the problem is that (in
contrast to R_w) $I(\lambda, w\mu)$ is not constant on conjugacy
classes. It is therefore less natural to average $I(\lambda, w\mu)$
over characters of $W(\mu)$.

A better analogue of the Steinberg representation is
$I(0,0)$ (called $I(0)$ in (7.18)). It has the character
formula

(7.28) $I(0,0) = |W|^{-1} \sum I(0, w \cdot 0),$

which is rather like the one in (7.9)(b). This suggests
that the unipotent representations of G might indeed

correspond roughly to representations of **W**, but with different representations of **W** requiring different infinitesimal characters. That turns out to be the case. It is not
yet convenient to state in detail what is true in this direction, but here is a weak version.

THEOREM 7.29 ([Barbasch-Vogan, 1985]. *Suppose* G *is a*
complex connected reductive algebraic group. To each of
certain double cells $\mathscr{C}$ *in* $\hat{W}$ *(Definition 7.10), one can*
associate a unique antidominant integral weight $\lambda = \lambda(\mathscr{C})$
in $(\mathfrak{h}_0)^*$ *(notation 7.17). Assume that* $\mathscr{C}$ *is such a cell.*
For σ *in* $\mathscr{C}$, *define*

$$R_\sigma = |W_\lambda|^{-1} \sum_{w \in W} tr(\sigma(w)) I(\lambda, w\mu).$$

   a) *The virtual representation* R_σ *is zero unless* σ
is of the form $\sigma(x)$, *for some* x *in* $A(\mathscr{C})$ *(notation*
7.13). In this case we write R_x *for* R_σ.

   b) *For each irreducible representation* π *of* $A(\mathscr{C})$,
there is an irreducible $(\mathfrak{g}, K)$-*module* $X(\pi)$. *In the*
Grothendieck group,

$$X(\pi) = |A(\mathscr{C})|^{-1} \sum_{x \in A(\mathscr{C})} tr(\pi(x)) R_x.$$

   c) *The formula in* (b) *may be inverted to give*

$$R_x = \sum_{\pi \in A(\mathscr{C})^{\wedge}} tr(\pi(x)) X(\pi).$$

A complete statement (saying when $\lambda(\mathscr{C})$ exists and how to construct it) will be given in Proposition 8.34. The cell of the sign representation will be attached to the weight $-\rho$, and that of the trivial representation to 0. This theorem therefore provides a common generalization of the irreducibility of $X(0)$ (cf. (7.18)), and Proposition 7.24.

Theorem 7.29 should be compared with Corollary 7.16. It is of the same general form, but substantially sharper. What happens is that the set of representations under consideration is much smaller in the complex case; so the formula analogous to Theorem 7.14 (which is Theorem 7.29(c)) can be inverted.

Theorem 7.29 has at least three serious faults. First, it gives no hint of how to realize the representation $X(\pi)$. Second, it treats only complex groups. Third, it does not say whether $X(\pi)$ is unitary. Barbasch has made enormous progress on resolving the last of these problems (see Chapter 8). The first two are nearly untouched.

Chapter 8

LANGLANDS' PRINCIPLE OF FUNCTORIALITY

AND UNIPOTENT REPRESENTATIONS

Langlands' principle of functoriality is a philosophy
about when automorphic forms on two different reductive
groups ought to have something to do with each other. It is
more properly the object of a lifetime of devoted study than
of a few pages of generalities. I cannot hope to do justice
to its motivation, to what is known to be true about it, or
even to what is known to be false. (Lack of space is a sufficient excuse on all counts, and my expertise need not be
considered.) But because we are looking only for guidance
in finding a good definition of unipotent representations,
we can afford to be careless about almost everything.

If F is a finite group, then the number of irreducible (complex) representations of F is equal to the number
of conjugacy classes in F. Only under special conditions

can one expect to find a natural bijection between these two
sets, however. If F is finite abelian, then we could define its dual group to be

(8.1)(a) $\qquad\qquad {}^{d}F = \mathrm{Hom}(F, \mathbb{C}^{x})$.

This is almost by definition the same as the group $\hat{F}$ of
characters of F, introduced in Lemma 1.2. We prefer to say
that there is a bijection

(8.1)(b)   {elements of ${}^{d}F$} $\leftrightarrow$ {characters of F}.

Here is another example of the same flavor, a little
closer to Langlands' basic idea. Let T be a compact connected abelian Lie group. We will summarize these assump-
tions by saying that T is a (compact) torus. Define

(8.2)(a) $\qquad\qquad X_{*}(T) = 2\pi i \cdot \ker(\exp)$,

a lattice in the real subspace $i t_{0}$ of the complexified Lie
algebra t of T. If we identify the unit circle $\mathbb{T}$ with
$2\pi i \mathbb{R}/2\pi i \mathbb{Z}$, then

(8.2)(b) $\qquad\qquad X_{*}(\mathbb{T}) = \mathbb{Z}$.

An easy argument now provides a natural isomorphism

(8.2)(c) $\qquad\qquad X_{*}(T) \cong \mathrm{Hom}(\mathbb{T}, T)$.

Because of this, we call $X_{*}(T)$ the lattice of *integral
one-parameter subgroups* of T. Using the identification
(8.2)(c), we can recover T from $X_{*}(T)$, by means of the
natural isomorphism

(8.2)(d) $\qquad\qquad T \cong X_{*}(T) \otimes_{\mathbb{Z}} \mathbb{T}$;

the map from right to left sends $h \otimes z$ to $h(z)$.

Put

(8.2)(e) $\qquad X^*(T) = \{\lambda \in \mathfrak{t}^* \mid \lambda(X_*(T)) \subset \mathbb{Z}\}$

$$\cong \text{Hom}_{\mathbb{Z}}(X_*(T), \mathbb{Z}),$$

the dual lattice to $X_*(T)$. This is the lattice Λ of Lemma 1.2, which is equal to $\Lambda_{\mathbb{C}}$ in this case. By Lemma 1.2(b) or (c),

(8.2)(f) $\qquad X^*(T) \cong \text{Hom}(T, \mathbb{T})$

$$\cong \hat{T}.$$

The duality between the lattices $X_*(T)$ and $X^*(T)$ suggests that we define the *dual torus* to T as

(8.3)(a) $\qquad {}^d T = X^*(T) \otimes_{\mathbb{Z}} \mathbb{T}.$

Then there are natural isomorphisms

$$X_*({}^d T) \cong X^*(T)$$

(8.3)(b) $\qquad X^*({}^d T) \cong X_*(T)$

$${}^d({}^d T) \cong T.$$

A more pedestrian form of the first two isomorphisms is

(8.3)(c) $\qquad {}^d \mathfrak{t} \cong \mathfrak{t}^*.$

PROPOSITION 8.4. *Let T be a compact torus, and ${}^d T$ the dual torus. Then there is a natural bijection between the characters of T (that is, homomorphisms from T to the unit circle $\mathbb{T}$) and integral one-parameter subgroups of ${}^d T$ (that is, homomorphisms from $\mathbb{T}$ to ${}^d T$).*

This is just a reformulation of (8.3)(b).

As an immediate consequence, we get a first hint at Langlands' functoriality principle.

COROLLARY 8.5. *Suppose* S *and* T *are compact tori. Assume that we are given a mapping*

$$\omega: {}^{d}S \rightarrow {}^{d}T$$

between the dual tori. Then there is a natural map (called transfer)

$$\text{tr}(\omega): \hat{S} \rightarrow \hat{T}$$

taking characters of S *to characters of* T. *That is, the set of characters of a torus is a covariant functor of the dual torus.*

For a character of S gives a map of $\mathbb{T}$ into ${}^{d}S$, and thus (by composition with ω) a map of $\mathbb{T}$ into ${}^{d}T$. This last map corresponds to a character of T.

This result should not be taken too seriously as it stands. The map ω actually induces a map

(8.6) $${}^{d}\omega: T \rightarrow S;$$

so a character of S induces one of T by composition with ${}^{d}\omega$. This covariant dependence of the group of characters on

the dual torus is therefore really just the contravariant
dependence on T.

Next, suppose K is a compact connected Lie group.
Fix a maximal torus T in K. Write R for the set
$\Delta(\mathfrak{k},\mathfrak{t})$ of roots of $\mathfrak{t}$ in $\mathfrak{k}$. By (1.8),

(8.7)(a) $R \subset X^*(T)$.

Each root α gives rise to a subgroup of K locally isomorphic to SU(2) (the two by two unitary matrices of
determinant one); that is, to a map

(8.7)(b) $\psi_\alpha : SU(2) \to G$.

Define a map $\alpha^\vee$ from $\mathbb{T}$ to T by

(8.7)(c) $\alpha^\vee(z) = \psi_\alpha(d(z,z^{-1}))$;

here d denotes the diagonal matrix with the indicated entries. By (8.2)(c), we can regard $\alpha^\vee$ as an element of
$X_*(T)$, and so as an element of $\mathfrak{t}$. If we identify $\mathfrak{t}$ and
$\mathfrak{t}^*$ using an invariant bilinear form as usual, then

(8.7)(d) $\alpha^\vee \leftrightarrow 2\alpha/(\alpha,\alpha)$.

We call $\alpha^\vee$ the *coroot corresponding to* α. Put

(8.7)(e) $R^\vee = \{\alpha^\vee \mid \alpha \in R\}$

                              $\subset X_*(T)$.

When more precision is necessary, we will write $\Delta^\vee(\mathfrak{k},\mathfrak{t})$
instead of $R^\vee$.

The quadruple $(X^*(T),R,X_*(T),R^\vee)$ is called a *root datum* for K. By the conjugacy of maximal tori in K, the root datum is unique up to isomorphism.

Definition 8.8 (Grothendieck–Demazure; see [Springer, 1979]). An *abstract root datum* is a quadruple $(L,\Phi,L^\vee,\Phi^\vee)$, subject to the conditions (i) – (iv) below. First, L and $L^\vee$ are assumed to be finitely generated free abelian groups, with

     i) $L^\vee \cong \mathrm{Hom}(L,\mathbb{Z})$.

The resulting pairing from $L \times L^\vee$ into $\mathbb{Z}$ is written $\langle\ ,\ \rangle$. Next, Φ and $\Phi^\vee$ are assumed to be finite subsets of L and $L^\vee$, respectively. They are assumed to be in one-to-one correspondence, by

     ii) $\alpha \leftrightarrow \alpha^\vee$.

Assume that

     iii) for all α in Φ, $\langle\alpha,\alpha^\vee\rangle = 2$.

If α belongs to Φ, define an endomorphism s_α of L by

$$s_\alpha(x) = x - \langle x,\alpha^\vee\rangle\alpha.$$

The endomorphism $s_{\alpha^\vee}$ of $L^\vee$ is defined similarly. Because of (iii), s_α and $s_{\alpha^\vee}$ are involutions. Assume that

     iv) for all α in Φ, $s_\alpha(\Phi) \subset \Phi$, and $s_{\alpha^\vee}(\Phi^\vee) \subset \Phi^\vee$.

A root datum is said to be *reduced* if it has the additional property

v) if $\alpha \in \Phi$, then $2\alpha \notin \Phi$.

This turns out to imply the corresponding property for $\Phi^\vee$.

If $\Psi = (L, \Phi, L^\vee, \Phi^\vee)$ is a (reduced) root datum, then the symmetry of the axioms implies that $^d\Psi = (L, \Phi, L, \Phi)$ is as well; it is called the *dual root datum* to Ψ.

Here is the Grothendieck–Demazure formulation of the standard structure theory for compact connected Lie groups.

PROPOSITION 8.9. *Suppose* K *is a compact connected Lie group. Then the root datum* Ψ *of* K *(defined in (8.7)) is an abstract reduced root datum. Suppose* K' *is another compact connected Lie group, with root datum* Ψ'. *Then* K *is isomorphic to* K' *if and only if* Ψ *is isomorphic to* Ψ'.

Conversely, if Ψ *is an abstract reduced root datum, then there is a compact connected Lie group* K *with root datum isomorphic to* Ψ.

Suppose K is a compact connected Lie group. A *dual group* for K is by definition a compact connected Lie group ^dK with root datum dual to that of K. To say what this means more concretely, fix maximal tori T for K and T' for ^dK. First, there should be given a distinguished class of isomorphisms

(8.10)(a) $\xi: {}^{d}T \to T'$

from the dual torus of T (cf. (8.3)) to T'. Any two of

these isomorphisms should differ by an element of the Weyl

group of T in K. Fix such a ξ. It will induce isomor-

phisms

(8.10)(b) $\xi: X^{*}(T) \to X_{*}(T')$

                       $\xi: X^{*}(T') \to X_{*}(T)$

(because of (8.3)(b)). Then the remaining condition is

(8.10)(c) $\xi(\Delta(\check{t},t)) = \Delta^{\vee}({}^{d}\check{t},t')$

                   $\xi(\Delta({}^{d}\check{t},t')) = \Delta^{\vee}(\check{t},t)$.

The notion of dual group for a non-abelian compact Lie

group is considerably more subtle than the one for tori.

The dual group ${}^{d}K$ exists only because of an abstract exis-

tence result (Proposition 8.9), and not because of a simple

construction like (8.3)(a). One consequence of this is that

duality is no longer a contravariant functor: a map between

compact groups does not always give rise to a map in the

other direction on dual groups.

Here are some examples of pairs $(K, {}^{d}K)$; we leave to

the reader the task of computing the root datum in each

case.

$$(U(n),U(n))$$

(8.11) $$(SU(n),PSU(n))$$

$$(SO(2n),SO(2n))$$

$$(SO(2n+1),Sp(n))$$

LEMMA 8.12. *Suppose* K *is a compact connected Lie group,
and* ^dK *is a dual group. Fix maximal tori* T *and* T' *for*
K *and* ^dK, *respectively. Then the Weyl groups* W(K,T) *and*
$W(^dK,T')$ *(Definition 1.28) are isomorphic, by an isomorphism
unique up to inner automorphisms.*

Proof. We can regard W(K,T) as acting on $X_*(T)$. A basic
fact about compact groups is that W(K,T) is generated by
the endomorphisms s_α defined in Definition 8.8. Because
of the duality between $X_*(T)$ and $X^*(T)$, it makes sense to
talk about the transpose of s_α, as an automorphism of
$X^*(T)$. An easy calculation from the definition gives

$$^t s_\alpha = s_\alpha.$$

It follows that W(K,T) is isomorphic to the group of automorphisms of $X^*(T)$ generated by the s_α, by the map

$$w \to {}^t w^{-1}.$$

On the other hand, the isomorphisms (8.10)(b) and (c) show

that this latter group is isomorphic to $W(^dK,T')$, by an iso-

morphism unique up to inner automorphism.   □

Any torus in ^dK is contained in a maximal torus. Com-

bining Proposition 8.4, Lemma 8.12, and Theorem 1.30, we

arrive at the following result.

PROPOSITION 8.13.   *Let* K *be a compact connected Lie group*

and ^dK *a dual group (cf. (8.10)). Then there is a natural*

bijection between the irreducible representations of K *and*

the ^dK-*conjugacy classes of homomorphisms from* $\mathbb{T}$ *to* ^dK.

COROLLARY 8.14.   *Suppose* K *and* H *are compact connected*

Lie groups. Assume that we are given a mapping

$$\omega: {}^dH \to {}^dK$$

between the dual groups. Then there is a map from irredu-

cible representations of H *to irreducible representations*

of K.

We have not given the induced map in Corollary 8.14 a name,

because it is not a good transfer. The problem is that it

is based on Theorem 1.30: it essentially parametrizes repre-

sentations by their highest weights. As we have seen

already in the Weyl character formula (Theorem 1.37), this

is not a good parametrization for the purpose of harmonic
analysis. We could fix the problem here, but the effort
would be wasted; there are no interesting transfers of the
kind we want between compact groups. We therefore proceed
directly to the next level of (forgive me) complexity.

Suppose now that G is a connected complex reductive
algebraic group, and that K is a maximal compact subgroup.
We use the notation (7.17). Recall that H is a Cartan subgroup of G.

THEOREM 8.15 (Zhelobenko; see [Duflo, 1975]). *Suppose* G
is a connected complex reductive algebraic group, H *is a*
Cartan subgroup, and W *is the Weyl group of* H *in* G.
Then there is a bijection

$$\hat{H}/W \leftrightarrow \hat{G}$$

from the set of W-*conjugacy classes of characters of* H,
onto the set of irreducible admissible representations of
G. *It may be defined as follows.* *Choose* H *as in (7.17).*
Fix a character of H, *associated to a pair* (λ, μ) *of ele-*
ments of $(\mathfrak{h}_0)^*$ *(Lemma 7.20). Then the corresponding repre-*
sentation of G *is* $J(\lambda, \mu)$ *(Definition 7.21).*

This is a rather serious result. We will not discuss the
proof (but see Theorem 13.1).

We have implicitly fixed already a maximal torus T in K. Fix a dual group ^dK of K (cf. (8.10)), and a maximal torus in ^dK. We will use a choice of the isomorphism in (8.10)(a) to identify this latter torus with ^dT, and write

(8.16)(a) $^dT \subset {}^dK$.

Choose a complexification ^dG of ^dK; this is just a complex connected reductive group having ^dK as a maximal compact subgroup. We call ^dG a *dual group* for G. Put

(8.16)(b) ^dH = centralizer of ^dT in ^dG,

a Cartan subgroup of ^dG. It is a complexification of ^dT, and so a dual group of H, as the notation indicates. Lemma 8.12 allows us to fix an identification of the Weyl groups for G and ^dG:

(8.16)(c) $W(G,H) \cong W(^dG, {}^dH)$.

Finally, (8.3)(c) gives

(8.16)(d) $^d\mathfrak{h} \cong \mathfrak{h}^*$.

The examples in (8.11) can be translated immediately into examples of pairs $(G, {}^dG)$. The most important of these is the pair $(GL(n,\mathbb{C}), GL(n,\mathbb{C}))$.

LEMMA 8.17 (Langlands). *Suppose T is a compact torus and H its complexification. Write ^dT for the dual torus (cf. (8.3)(a)) and ^dH for its complexification. Then there is* *a natural bijection from characters of* H *to homomorphisms of* $\mathbb{C}^\times$ *into* ^{d}H.

This is elementary. (Langlands' contribution was to notice that the fact might be interesting, and to generalize it to tori over any local field).

Definition 8.18. A continuous homomorphism ϕ of $\mathbb{C}^\times$ into a reductive group is called *quasi-admissible* if $\phi(\mathbb{C}^\times)$ consists of semisimple elements.

Combining Lemma 8.17 with the conjugacy of Cartan subgroups gives the following reformulation of Theorem 8.15.

THEOREM 8.19. (*The Langlands classification for complex groups*). *Suppose* G *is a complex connected reductive algebraic group, and* ^{d}G *is a dual group for* G *(cf. (8.16)). Then there is a natural bijection between the irreducible admissible representations of* G *and the* ^{d}G- *conjugacy classes of quasi-admissible homomorphisms of* $\mathbb{C}^\times$ *into* ^{d}G *(Definition 8.18).*

This theorem gives rise at last to interesting cases of "Langlands functoriality."

COROLLARY 8.20. *Suppose* G *and* G *are connected complex reductive algebraic groups. Assume that we are given a holomorphic homomorphism*

$$\omega: {}^d S \to {}^d G$$

between the dual groups (cf. (8.16)). Then there is a natural map (called transfer)

$$tr(\omega): \hat{S} \to \hat{G}$$

from irreducible admissible representations of S *to irreducible admissible representations of* G.

Langlands' principle of functoriality asserts that the essentially formal correspondences of Corollary 8.20 can have analytic content. For example, if the representation π of S appears in a space of square-integrable automorphic forms for S, then one might expect $tr(\omega)(\pi)$ to appear in such a space for G. It is a difficult matter to find precise and correct statements of this form, and I do not intend to try. The idea that we want to extract from Langlands' principle is that if π is a unitary representation of S, then $tr(\omega)(\pi)$ is a good candidate for a uni-

tary representation of G. (Any representation appearing in
a space of square-integrable automorphic forms is necessarily unitary.)

To see that this is a reasonable idea, consider the special case when S is equal to the Cartan subgroup H of
G. The construction of dual groups in (8.16) provides an
inclusion

(8.21)(a) $\omega: {}^d H \to {}^d G.$

The corresponding transfer is defined as follows. An irreducible admissible representation of H is a character, say
χ. Extend χ to the Borel subgroup HN, by making it trivial on N. Form the induced representation $I(\chi)$ (Definition 7.21). Let $J(\chi)$ be the irreducible subquotient of
$I(\chi)$ containing the representation of K of extremal
weight $\chi|_T$ (Definition 7.21). Then

(8.21)(b) $\mathrm{tr}(\omega)(\chi) = J(\chi),$

as one sees by tracing through the definitions. If χ is a
unitary character, then so is $I(\chi)$ (Theorem 3.19); so
$J(\chi)$ is unitary as well.

On the other hand, transfer does not always preserve
unitarity. For example, let G be $GL(3,\mathbb{C})$. By the remark
after (8.16), the dual group ${}^d G$ is again $GL(3,\mathbb{C})$. Let S
be $PGL(2,\mathbb{C})$, so that ${}^d S$ is $SL(2,\mathbb{C})$. The adjoint repre-

sentation of $^d S$ on its (three-dimensional) Lie algebra

gives a homomorphism

(8.22)(a) $\omega: {}^d S \to {}^d G.$

We want to compute the transfer on Stein complementary

series for $SL(2,\mathbb{C})$. Write $C(\sigma)$ for the restriction to

$SL(2)$ of the complementary series $C_2(k,t{:}\sigma)$ defined in

Theorem 4.23 for $GL(2)$; this is independent of the para-

meters k and t. This is a unitary representation for σ

smaller than 1. Define

$$\rho_\mathbb{C} = (\rho,\rho)$$

as in (7.22)(d); we may use a superscript to specify the

group. A calculation shows that

(8.22)(b) $C(\sigma) = J(\sigma\rho^S, \sigma\rho^S)$

(8.22)(c) $tr(\omega)(C(\sigma)) = J(\sigma\rho^G, \sigma\rho^G).$

It turns out that the representations of $GL(3,\mathbb{C})$ appearing

on the right are unitary only for σ less than or equal to

1/2. Transfer therefore fails to preserve unitarity.

The problem here is not with the functoriality princi-

ple, but with our abuse of it. The motivation to look for

preservation of unitarity came from the preservation of auto-

morphic forms. The Ramanujan–Petersson conjecture, however,

asserts that the representations $C(\sigma)$ do not appear on

nice spaces of square-integrable automorphic forms. Functoriality therefore should not encourage us to believe that they will transfer to interesting representations.

CONJECTURE 8.23 (cf. [Arthur, 1983]). *Suppose* G *and* S *are complex connected reductive algebraic groups, and*

$$\omega: {}^{d}S \to {}^{d}G$$

is a holomorphic homomorphism. Let χ *be a unitary character of* S. *Then the representation* $tr(\omega)(\chi)$ *(Corollary 8.20) is unitary.*

Arthur makes a far more detailed analysis of the conjectural role of these representations in the theory of automorphic forms. We will not formulate here Langlands' analogue of Theorem 8.19 for real algebraic groups. Once that is done, there is no difficulty in extending Corollary 8.20, and Conjecture 8.23; we refer to [Arthur, 1983] for more information.

Before discussing how to compute the representations in Arthur's conjecture, we present the most persuasive evidence of its validity now available.

THEOREM 8.24 (Barbasch). *Conjecture 8.23 is true if* G *is a (complex) classical group.*

This theorem produces many previously unknown unitary representations for almost all the classical groups except GL(n).
The proof is elegant in structure but very difficult to
carry out. The idea may well apply to the case of real
classical groups, but their character theory is not yet
sufficiently developed to attempt this.

We turn now to the problem of identifying the representations in Conjecture 8.23.

Definition 8.25 (cf. [Barbasch-Vogan, 1985]). Suppose G
is a complex connected reductive algebraic group. A representation π of G is called *spherical special unipotent*
if it is of the form $\mathrm{tr}(\omega)(\mathbb{C})$, for some ω as in Corollary
8.20. That is, we require π to be the transfer of the
trivial representation of some other group S.

Suppose S is PGL(2,$\mathbb{C}$), so that ^{d}S is SL(2,$\mathbb{C}$).
Fix a holomorphic homomorphism

(8.26)(a) $\psi\colon$ SL(2,$\mathbb{C}$) $\to$ ^{d}G.

Define

(8.26)(b) $H_{\psi} = d\psi \begin{bmatrix} 1 & 0 \\ 0 & -1 \end{bmatrix}$.

After replacing ψ by a conjugate under ^{d}G, we can and do assume that H_ψ is a dominant element of $^d\mathfrak{h}_0$. Under the identification (8.16)(d), H_ψ therefore corresponds to an element of $(\mathfrak{h}_0)^*$. Define

$$(8.26)(c) \qquad\qquad \lambda_\psi = \tfrac{1}{2}H_\psi,$$

regarded as an element of $(\mathfrak{h}_0)^*$. An important example is the case when G is itself equal to $PGL(2,\mathbb{C})$, and ψ is the identity. Then (in the notation of (7.22) or (8.22)),

$$(8.26)(d) \qquad\qquad \lambda_\psi = \rho^{PGL(2)}.$$

PROPOSITION 8.27. *Use the notation just described.*

   a) The trivial representation $\mathbb{C}$ of G is $J(\rho,\rho)$.

   b) The transfer (via ψ) from $PGL(2)$ to G of the trivial representation is $J(\lambda_\psi,\lambda_\psi)$:

$$tr(\psi)(\mathbb{C}) = J(\lambda_\psi,\lambda_\psi).$$

This representation of G contains the trivial representation of K.

   c) There is a homomorphism ψ_p from $SL(2,\mathbb{C})$ to ^dG, with the property that

$$\lambda_{\psi_p} = \rho.$$

   d) The transfer via ψ_p of the trivial representation of $PSL(2)$ to G is the trivial representation of G.

Proof. By (7.19), and (7.23), the trivial representation is a subrepresentation of $I(-\rho,-\rho)$ (appearing as the space of constant functions on G/B). By Lemma 7.25, the trivial representation is therefore a composition factor of $I(\rho,\rho)$. Since the trivial representation of G contains the trivial representation of K, which has highest weight

$$0 = \rho - \rho,$$

part (a) follows from Definition 7.21. Part (b) now follows from (a) (applied to the group $PGL(2,\mathbb{C})$) and the definition of transfer. Part (c) is due to Dynkin; a nice account is included in [Kostant, 1959]. In light of (a) (for G this time), part (d) is a special case of (b). $\square$.

COROLLARY 8.28. *The spherical special unipotent representations of G are those of the form* $J(\lambda_\psi,\lambda_\psi)$ *(Definition 7.21), with* λ_ψ *constructed from a homomorphism of* $SL(2,\mathbb{C})$ *into* ^dG *as in (8.26). Each contains the trivial representation of* K.

Corollary 8.28 says that the spherical special unipotent representations may be obtained from a far less general kind of functoriality than Definition 8.25 would suggest. The homomorphisms of $SL(2)$ into a reductive group G were completely determined by Dynkin; there are only finitely

many possibilities, up to conjugation in G. (If G is a

classical group, the question amounts to determining the set

of representations of SL(2) of a given dimension that pre-

serve some kind of bilinear form.)

Next, we will see how to construct the rest of the

representations of Conjecture 8.23.

PROPOSITION 8.29. *Suppose* G *is a complex connected reduc-*

tive algebraic group and π *is a representation of* G *as*

in Conjecture 8.23. Then there are

$\quad$ i) *a parabolic subgroup* $P = LN$ *of* G;

$\quad$ ii) *a spherical special unipotent representation* τ

of L; *and*

$\quad$ iii) *a unitary character* ξ *of* L,

with the property that the representation π *is contained*

in

$$\text{Ind}_P^G(\tau \otimes \xi).$$

The proof is easy, but we will omit it. There is in fact a

canonical choice for the conjugacy class (L, τ, ξ).

The induced representations in Proposition 8.29 can be

reducible. Since our goal is to find unitary representa-

tions, it is not reasonable to neglect the other constitu-

ents of $\text{Ind}(\tau \otimes \xi)$ (other than the one arising in Conjec-

ture 8.23). This suggests that the result of the transfer
should be, not the single representation provided by the
formal correspondence of Corollary 8.20, but some finite
family of representations. Arthur arrived at the same conclusion on the basis of much more sophisticated ideas about
the functoriality principle (the notion of "stability").
Here is a version of his idea.

CONJECTURE 8.30. *Suppose we are in the setting of Conjecture 8.23. Define*

   i) $Z(\omega)$ = *centralizer in* ^dG *of* $\omega(^dS)$.

*(This is a reductive algebraic group, but it need not be
connected.) Write*

   ii) $C(\omega) = Z(\omega)/Z_0(\omega)$,

the group of connected components of $Z(\omega)$. *Then there are
a certain quotient* $A(\omega)$ *of* $C(\omega)$, *and a finite family*
$\mathcal{F}(\omega,\chi)$ *of irreducible unitary representations of* G *associated to* ω *and the fixed unitary character* χ *of* S.
These have the following properties.

   a) The representation $tr(\omega)(\chi)$ *belongs to* $\mathcal{F}(\omega,\chi)$.

   b) There is a surjection from $\mathcal{F}$ *onto the group of
characters of* $A(\omega)$; *and* $tr(\omega)(\chi)$ *maps to the trivial
character.*

Arthur also requires that the surjection in (b) relate
nicely to the theory of endoscopic groups. Just as in the
case of Conjecture 8.23, there is a directly analogous formulation for real algebraic groups. The experts may notice
that Conjecture 8.30 is stated under much more general hypotheses than those of [Arthur, 1983], and that the conclusion
is weaker. (In his setting, Arthur asked for bijectivity in
(b) above.) It seems that Arthur's formulation is too
strong. Nevertheless, it is precisely right in an important
special case.

THEOREM 8.31 ([Barbasch-Vogan, 1985]). *Suppose* G *is a*
connected reductive algebraic group, and
$$\psi\colon SL(2,\mathbb{C}) \to {}^{d}G$$
is a holomorphic map. Define $C(\psi)$ *as in Conjecture 8.30,*
and let $A(\psi)$ *be Lusztig's canonical quotient of* $C(\psi)$
([Lusztig, 1984], 13.1, or [Barbasch-Vogan, 1985], 4.4).
Let π *be any irreducible representation of the (finite)*
group $A(\psi)$. *Then there is an irreducible* $(\mathfrak{g},K)$-*module*
$$X(\psi,\pi)$$
naturally attached to π. *For* π *equal to the trivial*
representation of $A(\psi)$, *we have*
$$X(\psi,1) = tr(\psi)(\mathbb{C}) = J(\lambda_{\psi},\lambda_{\psi}).$$

We have been a little imprecise in the matter of defining

$A(\psi)$ here.   In fact, one has to work inside the centralizer

of $\psi(-I)$, and not inside all of ^dG.

The requirements of Arthur's conjecture related to endo-

scopy are satisfied by the correspondence described by

Theorem 8.31.   What is missing from [Barbasch–Vogan, 1985]

is a proof that the representations $X(\psi,\pi)$ are unitary.

Barbasch has since proved this for the classical groups;

this is essentially Theorem 8.24.

Definition 8.32 ([Barbasch–Vogan, 1985]).   Suppose G is a

complex connected reductive algebraic group.   A representa-

tion of G is called *special unipotent* if it is of the form

$X(\psi,\pi)$ (cf. Theorem 8.31) for some ψ mapping $SL(2,\mathbb{C})$ to

^dG.

We will conclude this chapter by explaining the rela-

tionship between Theorems 7.29 and 8.31.   Fix a map ψ and

other notation as in (8.26). Assume that

(8.33)(a) $\psi(-I) \in$ center of ^dG.

Here   I   refers to the identity matrix in $SL(2,\mathbb{C})$.   This is

equivalent to

(8.33)(b) λ_ψ is an integral weight.

Put

(8.33)(c) $\qquad W_\psi = \{w \in W | \ w\lambda_\psi = \lambda_\psi\}.$

Let p_ψ be the product of the positive roots orthogonal to λ_ψ, a polynomial function $\mathfrak{h}_0$. If w belongs to W_ψ, then

$$w \cdot p_\psi = \det(w)p_\psi.$$

The translates of p_ψ under W span a vector space $(V_\psi)'$; the representation $(\sigma_\psi)'$ of W on it is irreducible. Put

(8.33)(d) $\qquad \sigma_\psi = (\sigma_\psi)' \otimes \mathrm{sgn};$

here sgn denotes the sign representation of W. Finally, write

(8.33)(e) $\qquad \mathfrak{C}_\psi$

for the double cell in $\hat{W}$ containing σ_ψ (Definition 7.10).

PROPOSITION 8.34. *In the setting above, the double cell* $\mathfrak{C}_\psi$ *determines the weight* λ_ψ *up to conjugacy. The group* $A(\psi)$ *of Theorem 8.31 is canonically isomorphic to the group* $A(\mathfrak{C}_\psi)$ *attached to the double cell (Theorem 7.12).*

a) The Weyl group representation σ_ψ *is* $\sigma(e)$ *(notation (7.13)).*

b) Suppose π *is a character of* $A(\psi)$. *The* $(\mathfrak{g},K)$-*module* $X(\pi)$ *of Theorem 7.29 coincides with* $X(\psi,\pi)$ *(Theorem 8.31).*

This finally provides an (indirect) description of the

weight λ of Theorem 7.29 and the set of double cells aris-

ing there; the idea is to look at all possible ψ (subject

to (8.33)(a)) and see what turns up.

Chapter 9

PRIMITIVE IDEALS AND UNIPOTENT REPRESENTATIONS

In this chapter, we will outline a little of the theory
of primitive ideals in the enveloping algebra $U(\mathfrak{g})$ of a
reductive Lie algebra. As usual, we wish to go only far
enough to see how that theory impinges on unitary representation theory. A more complete (and less utilitarian) outline
may be found in [Joseph, 1983].

Definition 9.1. Suppose R is a (possibly non-commutative)
ring with 1, and M is a left R-module. The *annihilator*
of M is the two-sided ideal
$$\text{Ann } M = \{r \in R \mid rm = 0, \text{ all } m \in M\}.$$
M is called *simple* if it is not zero and it has no proper
submodules. An ideal in R is called (left) *primitive* if
it is the annihilator of a simple (left) module. Set

Prim R = {primitive ideals in R},

the (left) *primitive spectrum* of R.

Primitive ideals are in many respects the best generalization to non-commutative rings of maximal ideals in the commutative case. The following proposition gives some hint of
this.

PROPOSITION 9.2. *In the setting of Definition 9.1, any max-*
imal ideal is primitive. If R *is commutative, any primi-*
tive ideal is maximal.

Proof. Let I be a maximal ideal. By Zorn's lemma, there
is a maximal proper left ideal m containing I. It is
easy to check that the module R/m is simple and has annihilator I.

Conversely, suppose R is commutative, M is simple,
and m is a non-zero element of M. Let m be the annihilator of the element m. Then

$$\text{Ann}(R \cdot m) = m$$

$$R \cdot m = M$$

$$M \cong R/m.$$

(The first statement uses the commutativity of R; the
second uses the simplicity of M; and the third follows from

the second.) Because of the first two statements, we must show m is maximal. By the third, this is equivalent to the simplicity of M.   □.

Here is an assortment of useful technicalities.

Definition 9.3. Suppose R is a ring with 1 and I is an ideal in R. We say that I is *prime* if it is proper and the following condition holds: whenever J and J' are ideals containing I, and JJ' is contained in I, then either J or J' is equal to I. I is called *completely prime* if it is proper and the following condition holds: whenever r and s are elements of R, and rs belongs to I, then either r or s belongs to I.

R is called *prime* (respectively *completely prime*, *primitive*) if the zero ideal is prime (respectively completely prime, primitive). R is called *simple* if the zero ideal is maximal and proper.

PROPOSITION 9.4. *A maximal ideal is primitive; a primitive ideal is prime; and a completely prime ideal is prime. No other implications hold among these properties in general.*

If R *is commutative, completely prime is equivalent to prime, and primitive is equivalent to maximal.*

Proof. We know most of this from Proposition 9.2. Suppose
I is primitive; say I is the annihilator of the simple
module M. We want to show that I is prime. Suppose J
and J' contain I but neither is equal to I. Because
J' is strictly larger than the annihilator of M, J'M is
a non-zero submodule of M. Since M is simple, it coincides with M. Consequently J(J'M) is also non-zero.
Therefore JJ' does not annihilate M; so JJ' is not
contained in I, as we wished to show.

   To see that maximal does not imply completely prime,
take R to be the ring of n×n matrices and I to be the
zero ideal. Since R is simple, I is maximal, primitive,
and prime. But if n is at least two, there are non-zero
nilpotent elements in R; so I cannot be completely prime.

   The rest of the proposition will be left to the reader.
(The only non-trivial part is the construction of a non-
simple primitive ring.)   □.

   The example of the difference between prime and completely prime may be pushed to suggest a connection with
unitary representations.

PROPOSITION 9.5. *Suppose G is a noncompact connected simple Lie group, and F is a finite-dimensional irreducible representation of G.*

  a) F is unitary if and only if F has dimension 1.

  b) Ann(F) is a completely prime ideal if and only if F has dimension 1.

Proof. For (a), the commutator subgroup of G is all of G; so any one-dimensional representation is trivial, and therefore unitary. Conversely, suppose F is unitary. Then G maps into the (compact) unitary group of F. Because G is simple, it follows that either F is trivial or G is compact. We are assuming G is not compact; so (a) follows.

  For (b), the irreducibility of F guarantees that every endomorphism of F comes from the enveloping algebra of $\mathfrak{g}$:

$$U(\mathfrak{g})/\text{Ann } F \cong \text{End}(F).$$

We saw in the proof of Proposition 9.4 that End(F) is completely prime only when F is one-dimensional.   □.

  Inspired by a variety of much deeper examples, Joseph suggested that unitary representations should be closely connected with completely prime ideals, at least for complex groups. Here is a result along those lines.

PROPOSITION 9.6 ([Vogan, 1986a], Proposition 7.12). *Suppose* G *is a complex connected reductive algebraic group, and* X *is an irreducible unitary* $(\mathfrak{g},K)$*-module. Then* Ann(X) *is a completely prime primitive ideal in* $U(\mathfrak{g})$. *More precisely, write*

$$U(\mathfrak{g}) \cong U(\mathfrak{g}^L) \otimes U(\mathfrak{g}^R),$$

in accordance with (7.22). (Thus $\mathfrak{g}^L$ *and* $\mathfrak{g}^R$ *are each isomorphic to* $\mathfrak{g}_0$.) *Then*

$$\text{Ann}(X) = I^L \otimes U(\mathfrak{g}^R) + U(\mathfrak{g}^L) \otimes I^R$$

$$U(\mathfrak{g})/\text{Ann}(X) \cong (U(\mathfrak{g}^L)/I^L) \otimes (U(\mathfrak{g}^R)/I^R).$$

The ideals I^L *and* I^R *are each completely prime.*

The proof is quite easy but requires a little notation. We refer the reader to [Vogan, 1986a] for details.

Definition 9.7. Suppose G is a complex connected reductive algebraic group. Define $\mathfrak{g}^L$ and $\mathfrak{g}^R$ (both isomorphic to the Lie algebra $\mathfrak{g}_0$ of G) as in (7.22). If X is any $\mathfrak{g}$-module, set

$$\text{LAnn}(X) = U(\mathfrak{g}^L) \cap \text{Ann}(X)$$

$$\subset U(\mathfrak{g}^L) \cong U(\mathfrak{g}_0).$$

the *left annihilator* of X.   Similarly, we define the right
annihilator   RAnn(X).

Unfortunately, the analogue of Proposition 9.6 is false
for real groups (for example if G is compact).   It is true
for SL(n,ℝ), however, as a consequence of the classification of all unitary representations in that case.   This suggests a weak version that might be true in general.

CONJECTURE 9.8.   *Suppose* G *is a quasisplit real reductive*
algebraic group and X *is a unitary* (g,K)-*module of finite*
length.   *Assume that* X *is stable in Langlands' sense; that*
is (roughly speaking), that the distribution character of X
is constant on the intersection with G *of strongly regular*
semisimple conjugacy classes for the complexification G(ℂ).
Then U(g)/Ann(X) *has no non-zero nilpotents.*   *In particu-*
lar, if Ann(X) *is prime, then it is completely prime.*

Langlands' functoriality principle suggests that one might
be able to lift X to a representation of G(ℂ) in some
sense.   If this were possible, the conjecture might follow
from Proposition 9.6.

Proposition 9.6 produces completely prime primitive
ideals from unitary representations.   Our main interest is

in finding unitary representations, however; so we need to
know the extent to which Proposition 9.6 admits a converse.
The obvious converse (that every irreducible representation
whose annihilator is completely prime must be unitary) is
false for G equal to $\mathbb{C}^X$: the ideal must also be self-
adjoint in a certain sense. Here is a definition.

Definition 9.9. Suppose $\mathfrak{g}_0$ is a real Lie algebra with
complexification $\mathfrak{g}$. Write

$$u \to u^h$$

for the (conjugate linear) anti-automorphism of $U(\mathfrak{g})$ defined by

$$(X + iY)^h = -X + iY$$

for X and Y in $\mathfrak{g}_0$. If I is an ideal in $U(\mathfrak{g})$, then
its image I^h is as well. We say that I is *self-adjoint*
if I is equal to I^h.

Once this additional condition is imposed, the converse is
true for commutative G; but it fails for $SL(2,\mathbb{C})$, for
(easy) reasons connected with the complementary series.
Another of our interests is avoiding the problem of complementary series; so we would like a way to say that an ideal
has nothing to do with them. This requires a small digres-

sion on the analogue of Mackey induction (Definition 3.8)
for ideals.

Definition 9.10 (cf. [Dixmier, 1974]). Suppose $\mathfrak{g}$ is a
complex Lie algebra and $\mathfrak{h}$ is a subalgebra. Let J be an
ideal in $U(\mathfrak{h})$; say J is the annihilator of a representa-
tion V of $\mathfrak{h}$. Let δ be the modular character of $\mathfrak{g}/\mathfrak{h}$;
this is the one-dimensional representation of $\mathfrak{h}$ defined by

$$\delta(X) = [\mathrm{tr}\ \mathrm{ad}_{\mathfrak{g}}(X)] - [\mathrm{tr}\ \mathrm{ad}_{\mathfrak{h}}(X)].$$

Let V' be the tensor product of V with the character δ,
and J' the annihilator of V'. Put

$$W = \mathrm{pro}_{\mathfrak{h}}^{\mathfrak{g}}(V')$$

(Definition 6.12). The *ideal induced by* J is

$$I = \mathrm{Ann}(W);$$

we also write it as $\mathrm{pro}_{\mathfrak{h}}^{\mathfrak{g}}(J)$.

This definition differs very slightly from that in [Dixmier,
1974]. If G is a connected group with Lie algebra $\mathfrak{g}$,
then

(9.11) $\mathrm{pro}(J) = \{u \in U(\mathfrak{g}) \mid \mathrm{Ad}(g)(u) \in J'U(\mathfrak{g}), \text{ all g in G}\}.$

Dixmier's definition would have instead $U(\mathfrak{g})J'$. Ours is
more obviously compatible with group-theoretic induction
(cf. Proposition 6.13).

The next result is well-known; the proof outlined uses the ideas in [Conze, 1973].

LEMMA 9.12. *In the setting of Definition 9.10, suppose* J *is a completely prime ideal in* $U(\mathfrak{h})$. *Then the induced ideal* I *is a completely prime ideal in* $U(\mathfrak{g})$.

Sketch of proof. Fix a connected group G with Lie algebra $\mathfrak{g}$. Proposition 6.13 says that W looks like a space of sections of a vector bundle (on G/H). We want $U(\mathfrak{g})/I$ to look like a ring of differential operators on the bundle. Write A for the ring $U(\mathfrak{h})/J'$, and B for $U(\mathfrak{g})/I$. Because I is a two-sided ideal, it is invariant under the action ad of $\mathfrak{g}$, and so also under the action Ad of G. It follows that Ad is defined on B. Recall the order of vanishing filtration (Definition 6.12) of W. An endomorphism of W is said to be of *order* (at most) k *at the identity* if it maps W_k into W_0. Notice that the endomorphism given by the action of an element of $U_k(\mathfrak{g})$ has order at most k at the identity. An element b of B is said to be of *order* (at most) k if $Ad(g^{-1})(b)$ has order at most k at the identity, for all g in G. The filtration of W may be described as

$$W_m = \{\phi \in W | \ u\phi \in W_0, \ \text{all } u \in U_m(\mathfrak{g})\}.$$

Using this and the ad-invariance of order, one can check
that an element of order k maps W_m into W_{m-k}, for all
m. It follows that if u has order k, and v has order
j, then uv has order $k+j$.

The ring B therefore has an increasing filtration
$\{B_k\}$, compatible with (but not necessarily equal to) the one
induced by $\{U_k(\mathfrak{g})\}$. To prove that B is a completely
prime ring, it suffices to prove that $gr(B)$ is. For that
we need a symbol calculus. Notice first that

$$W_{m-1}/W_m \cong \operatorname{Hom}_{\mathbb{C}}(S^m(\mathfrak{g}/\mathfrak{h}), V')$$
$$\cong S^m((\mathfrak{g}/\mathfrak{h})^*) \otimes V';$$

we will call this space $gr(W)_m$. If T is an endomorphism
of W of order k at the identity, then T induces an
element

$$\sigma_k(T)(e) \in \operatorname{Hom}(gr(W)_k, gr(W)_0)$$
$$\cong S^k(\mathfrak{g}/\mathfrak{h}) \otimes \operatorname{Hom}(V', V'),$$

called the kth-order symbol of T at the identity. An
element u of $U(\mathfrak{g})$ of order k at the identity can be
written

$$u = \sum h_r x_r + \sum j_s' y_s,$$

with x_r in $U_k(\mathfrak{g})$, h_r in $U(\mathfrak{h})$, y_s in $U(\mathfrak{g})$, and j_s'
in J'. (This is a consequence of the Poincaré–Birkhoff–
Witt theorem, but it requires a little thought.) Then it
follows from the definition that

$$\sigma_k(u)(e) = \sum gr(x_r)a_r$$
$$\in S^k(\mathfrak{g}/\mathfrak{h}) \otimes A.$$

Here a_r is the image of h_r in A.

We now have a symbol map at the identity

$$\sigma_k(*)(e): gr_k B \to S^k(\mathfrak{g}/\mathfrak{h}) \otimes A.$$

If b belongs to B_k, the symbol of b is defined to be
the function on G with values in $S(\mathfrak{g}/\mathfrak{h}) \otimes A$, defined by

$$\sigma_k(b)(g) = \sigma_k(\mathrm{Ad}(g^{-1})(b))(e).$$

One can check that the symbol is a holomorphic function of
g. (For a fixed b, its values are in a finite-dimensional
subspace of $S^k(\mathfrak{g}/\mathfrak{h}) \otimes A$; so "holomorphic" makes sense.)
The symbol vanishes if and only if b belongs to B_{k+1}.
Finally, one can show that if u is in B_k, and T has
order j at the identity, then Tu has order $k+j$ at the
identity; and

$$\sigma_{k+j}(Tu)(e) = [\sigma_k(T)(e)][\sigma_j(u)(e)].$$

We have therefore embedded $gr(B)$ in the ring of holomor-
phic functions on G with values in $S(\mathfrak{g}/\mathfrak{h}) \otimes A$. This ring
clearly has no zero divisors.   □.

In the setting of Definition 9.7, assume that $P = MN$
is a parabolic subgroup of G, and that X_M is a Harish-
Chandra module for M, with left annihilator I_M. (We are

writing M instead of L for the Levi factor to avoid confusion with the L that means left.) Set

(9.13)(a) $$X = \text{Ind}_P^G(X_M),$$

a Harish-Chandra module for G, and

(9.13)(b) $$I = \text{LAnn}(X).$$

Then

(9.13)(c) $$I = \text{pro}_{\mathfrak{p}_0}^{\mathfrak{p}_0}(I_M)$$

(Definition 9.10), as is easy to check.

THEOREM 9.14 ([Joseph, 1980]). *Let* $\mathfrak{g}$ *be a complex semisimple Lie algebra. Then there are only finitely many completely prime primitive ideals in* $U(\mathfrak{g})$ *which are not
induced from any completely prime primitive ideal on a
proper parabolic subalgebra (in the sense of Definition
9.10).*

This is an extremely difficult result. Its proof does not
easily provide a list of the missing completely prime primitive ideals; in fact much of Joseph's subsequent work is
aimed at doing that.

Theorem 9.14 suggests (in conjunction with Proposition
9.6 and (9.13)(c)) the following conjecture.

(False) CONJECTURE 9.15. *Suppose* G *is a connected complex semisimple Lie group, and* X *is an irreducible Harish–Chandra module for* G. *Assume that* Ann(X) *is a self–adjoint completely prime primitive ideal, not induced from a proper parabolic subalgebra. Then* X *is unitary.*

Before considering what is wrong with this conjecture, let us consider what is right with it. The only obvious completely prime ideal that fails to be induced is the augmentation ideal $\mathfrak{g}_0 U(\mathfrak{g}_0)$. The only irreducible Harish–Chandra module of which it is the annihilator is the trivial representation, which is unitary. For $SL(n, \mathbb{C})$, Moeglin has proved that the augmentation ideal is the only non–induced completely prime ideal; so Conjecture 9.15 is true in that case.

The next simplest non–induced completely prime ideal is in the enveloping algebra of $\mathfrak{sp}(4, \mathbb{C})$ (type B_2 or C_2). It is the left annihilator I of the metaplectic representation (mentioned in the Interlude). In primitive ideal theory, I is called the Joseph ideal, because of its construction in [Joseph, 1976]. In fact, it can easily be shown that any irreducible Harish–Chandra module with left and right annihilators equal to   I   is a component of the

metaplectic representation, and therefore unitary.

The theory of special unipotent representations, de-

scribed in section 8, also fits well with Conjecture 9.15.

PROPOSITION 9.16 ([Barbasch-Vogan, 1985]). *In the setting*

of Theorem 8.31, we have (for fixed ψ)

$\{X(\psi,\pi)\}$ = $\{$Harish-Chandra modules $X|$ Ann(X) = Ann $X(\psi,1)\}$.

This is really the definition of special unipotent used in

[Barbasch-Vogan, 1985]. One should keep in mind that

$$X(\psi,1) = tr(\psi)(trivial)$$

(with the transfer tr defined by Corollary 8.20). Now

Theorem 8.24 becomes evidence for Conjecture 9.15 as well;

for most of the ideals involved here are not induced.

I do not know any counterexamples to Conjecture 9.15

for the classical groups. In type G_2 , however, there is a

problem. Joseph has shown that there are exactly two com-

pletely prime primitive ideals I_1 and I_2 in the envelop-

ing algebra, such that each quotient $U(g)/I_i$ has Gelfand-

Kirillov dimension 8 (see [Joseph, 1981] and [Vogan, 1986a],

section 5). These ideals are not induced. There are unique

irreducible Harish-Chandra modules X_1 and X_2 such that

$$\text{LAnn}(X_i) = \text{RAnn}(X_i) = I_i.$$

However, Duflo's results in [Duflo, 1979] show that only one

of these two representations (say X_1) is unitary. Accord-

ing to the Dixmier conjecture as formulated in [Vogan,

1986a], I_2 corresponds to a certain non-normal algebraic

variety, and I_1 to its normalization. This is a small and

subtle distinction, and it is hard to see how it can direct-

ly affect representation theory. (One might hope that the

ring $R_2 = U(g)/I_2$ is itself "non-normal" in some sense,

and that this causes the problem. Joseph's work suggests

that a reasonable definition of normal for a primitive

quotient R of $U(g)$ is that R should be the largest

Harish-Chandra module in $\text{Fract}(R)$. This condition is

satisfied for R_2, however.)

Without understanding the G_2 example, we can still

use it to guide a reformulation of Conjecture 9.15. We

begin by recalling from [Vogan, 1986a] a part of the Dixmier

conjecture alluded to above.

CONJECTURE 9.17. Suppose G is a connected reductive alge-

braic group, with Lie algebra g_0. Fix an orbit Y^o of G

on g_0^*, and write Y for its closure. Fix an irreducible

affine algebraic variety V, endowed with

    1) an algebraic action of G, and

2) a finite, G-equivariant morphism π from V onto Y.

Then there is canonically associated to V a completely prime primitive algebra A = A(V), endowed with

i) an algebraic action (called Ad) of G by automorphisms, and

ii) a G-equivariant (for the adjoint action on $U(g)$) algebra homomorphism ϕ from $U(g)$ to A, making A a finitely generated $U(g)$-module.

We require in addition that the differential ad of Ad be

$$ad(X)a = Xa - aX$$

for all a in A(V), and that A(V) be isomorphic as a G-module (but not as an algebra) to the ring of algebraic functions on V.

The Dixmier conjecture says that this association should provide a bijection from varieties satisfying (1) and (2), onto completely prime primitive algebras satisfying (i) and (ii).

The conditions in Conjecture 9.17 are not sufficient to specify A(V) uniquely in general; but for the non-induced cases, they probably suffice. The problem of finding a candidate for A(V) can be reduced to those cases. Many inter-

esting examples are known, but there are few good general
results.

To set the stage for an improvement on Conjecture 9.15,
we need some (conjectural) supplementary information about
the Dixmier correspondence. Algebraic geometry would suggest that the simple algebras should correspond to closed
orbits. This is very far from true; it is not at all clear
how to guess when A(V) is simple. Here is a possible sufficient condition, however.

CONJECTURE 9.18. In the setting of Conjecture 9.17, assume
that Y^o is a nilpotent orbit and that the variety V is
normal. Then the algebra A(V) is simple; so the kernel
I(V) of the homomorphism ϕ (Conjecture 9.17(ii)) is a
completely prime maximal ideal in U(g).

"Nilpotent" will be defined carefully after (10.19) below.
We call the primitive ideals I(V) *unipotent primitive
ideals*.

Here at last is a possible partial converse for Proposition 9.6.

CONJECTURE 9.19. Suppose G is a connected reductive algebraic group, Y^o is a nilpotent orbit in $(g_o)^*$, and Y is
the closure of Y^o. Fix an irreducible, normal, affine algebraic variety V, endowed with a G action and a finite
equivariant morphism onto Y. Let I(V) be the (completely
prime) maximal ideal of Conjecture 9.18. Let X be any
irreducible Harish-Chandra module having left and right annihilator equal to I(V). Then X is unitary.

"Definition" 9.20. Suppose G is a complex connected reductive algebraic group. An irreducible representation of G
is called unipotent if its left and right annihilators are
both equal to one of the ideals I(V) in Conjecture 9.19.

The quotation marks are there because I(V) has not
really been defined; so unipotent is not defined either.
For classical G, there is an explicit candidate for I(V)
(defined on a case-by-case basis); so the definition is complete. Since Barbasch has determined the unitary representations of classical complex groups, one could in principal
check Conjecture 9.19 in that case. This has not been done.
Definition 9.20 is not sufficiently precise to decide

whether special unipotent representations (Definition 8.32)

must be unipotent. Certainly this ought to be true.

We would like to understand the varieties V that

appear. In the setting of Conjecture 9.17, the fact that

coadjoint orbits are even-dimensional means that $Y-Y^o$ has

(complex) codimension two in Y. It follows that Y^o and

Y have the same fundamental group $\pi_1(Y)$. This group is

finite. By Zariski's Main Theorem, the normal varieties

mapping finitely onto Y are in one-to-one correspondence

with the coverings of Y^o, and hence with subgroups of

$\pi_1(Y)$. All of this can be done equivariantly; and we get

LEMMA 9.21. *Suppose* G *is a connected reductive algebraic*

group, Y^o *is an orbit in* $(g_0)^*$, *and* Y *is the closure of*

Y^o. *Fix a point* y *in* Y^o, *and put*

$$H_y = \textit{stabilizer of } y$$

$$H_0 = \textit{identity component of } H_y$$

$$\pi_1(G,Y) = H_y/H_0.$$

Write $\Pi(Y)$ *for the set of equivalence classes of irreduci-*

ble, normal, affine algebraic varieties V *endowed with the*

structures (1) and (2) of Conjecture 9.17. Then $\Pi(Y)$ *is*

in bijection with the set of conjugacy classes of subgroups

of $\pi_1(G,Y)$, *as follows. Fix such a covering* V, *and a pre-*

image v *of* y *in* V. Write H_v *for the stabilizer of* v
in G. *Then*

$$H_0 \subset H_v \subset H_y;$$

so

$$\pi_1(G,V) = H_v/H_0$$

may be regarded as a subgroup of $\pi_1(G,Y)$. *The bijection*
sends the equivalence class of V *to the conjugacy class of*
$\pi_1(G,V)$.

Example 9.22. Suppose G is Spin(9,$\mathbb{C}$) (the double cover
of G' = SO(9,$\mathbb{C}$)). We may regard $(\mathfrak{g}_0)^*$ as the space of
9×9 complex skew-symmetric matrices. Let Y^0 be the
coadjoint orbit consisting of nilpotent matrices with Jordan
blocks of size 5, 3, and 1. Then one can check that

$$\pi_1(G',Y) \cong (\mathbb{Z}/2\mathbb{Z}) \times (\mathbb{Z}/2\mathbb{Z})$$

$$\pi_1(G,Y) \cong (\mathbb{Z}/4\mathbb{Z}) \times (\mathbb{Z}/2\mathbb{Z}) \text{ (the dihedral group)}.$$

Write D for the dihedral group. Here is a description of
the ideals I(V) corresponding to various coverings of Y
(and hence to subgroups S of D). Each I(V) is a maximal ideal, and therefore (by an observation of Dixmier) is
determined by its intersection with the center of $U(\mathfrak{g}_0)$.
By Harish-Chandra's theorem (Theorem 6.4), this intersection
is determined by an element λ (or $\lambda(S)$) in the dual
$(\mathfrak{h}_0)^*$ of a Cartan subalgebra. There is a standard way to

identify $(\mathfrak{h}_0)^*$ with $\mathbb{C}^4$ (see [Humphreys, 1972]); so finally I(V) is determined by an element of $\mathbb{C}^4$, still written λ.

To Y itself (that is, to the subgroup D), we attach $(\frac{1}{2},\frac{1}{2},\frac{1}{2},0)$. This is one of the parameters λ_ψ of (8.26); so the (unique) corresponding unipotent representation is special unipotent. It is unitarily induced from the trivial representation on (the parabolic subgroup with Levi factor locally isomorphic to) $GL(1) \times GL(2) \times SO(3)$.

To one of the two $(\mathbb{Z}/2\mathbb{Z})^2$ subgroups of D, we attach $(\frac{1}{2},\frac{1}{2},\frac{1}{2},\frac{1}{2})$. This is again special unipotent; the two representations attached are the two constituents of the representation induced from the trivial representation on $GL(2) \times GL(2)$.

To the second $(\mathbb{Z}/2\mathbb{Z})^2$ subgroup, we attach $(1,\frac{1}{2},0,0)$. This is not special unipotent. Since $SO(5,\mathbb{C})$ is isomorphic to $Sp(4,\mathbb{C})$ modulo its center Z, the component of the metaplectic representation trivial on Z gives a unitary representation of $SO(5)$. The unipotent representations here are the two components of the representation induced from the metaplectic representation on $GL(1) \times GL(1) \times SO(5)$.

To one of the classes of non-normal $\mathbb{Z}/2\mathbb{Z}$ subgroups, we attach $\frac{1}{4}(3,3,1,1)$. This is not special. The unipotent representations are unitarily induced from two unipotent representations of GL(4), attached to the double cover of the nilpotent orbit with Jordan blocks of sizes 2 and 2.

To the other class of non-normal $\mathbb{Z}/2\mathbb{Z}$ subgroups, we attach $(1,\frac{1}{2},\frac{1}{4},\frac{1}{4})$. The two unipotent representations are unitarily induced from GL(2)×SO(5).

To the trivial subgroup of D – that is, to the universal cover of Y – we attach $(1,\frac{1}{2},\frac{1}{2},0)$. There are five unipotent representations with annihilator I(V): four complementary series induced from various characters of GL(1)×GL(2)×SO(3), and one induced from a character of GL(2)×GL(2). This latter representation is twice as large as the first four. In light of Theorem 8.31, this suggests that the five representations should be parametrized by the irreducible representations of D.

It is not completely clear what the weights attached to the two other subgroups of D ought to be. Possibly one simply uses again the weight attached to all of D. This would satisfy many of the formal requirements of Conjecture 9.17.

Several things emerge from this example. First, the algebras A(V) of Conjecture 9.17 do not vary nicely with V, for fixed Y. Second, the more obvious conjectures about how to parametrize the unipotent representations attached to

V   seem to fail: the representations are related in some way

to the character theory of the fundamental groups of Y   and

V, but not by a result as clean as Theorem 8.31.   Finally,

the weight λ attached to V   seems to increase slightly in

size with V.   This suggests that the weight attached to the

normalization of Y   (the smallest covering under consider-

ation) should be the smallest infinitesimal character admit-

ting a primitive ideal attached to the orbit Y^o.   This

weight is explicitly computable; an incomplete result in

that direction may be found in Proposition 5.10 of

[Barbasch-Vogan, 1985].

Chapter 10

THE ORBIT METHOD AND UNIPOTENT REPRESENTATIONS

To understand the origins of the method of coadjoint
orbits, we must return for a moment to the general setting
of the introduction. Suppose G is a nilpotent Lie group.
Then G has a rich supply of normal subgroups. As explained before Theorem 0.5, this makes it possible to
describe the representation theory of G in terms of that
of smaller groups; eventually one comes down to the case of
abelian groups and ordinary characters (Lemma 1.2). All of
this was well understood in the 1950's, thanks to the work
of Mackey and others. Unfortunately, the answers provided
by this method were a little difficult to understand syste-
matically and interpret. The simplest non-abelian nilpotent
group is the three-dimensional Heisenberg group (mentioned
and defined in the introduction). It has a family of characters parametrized by $\mathbb{R}^2$, and a family of infinite-

dimensional representations parametrized by $\mathbb{R}\setminus\{0\}$. This is certainly explicit, but it is only an answer; there is nothing compelling, enlightening, or beautiful about it.

In the 1960's, Kirillov and Kostant found a way of thinking about representations which overcomes these problems. Here is its first great success.

THEOREM 10.1 ([Kirillov, 1962]). *Suppose* G *is a connected, simply connected nilpotent Lie group. Write* $\mathfrak{g}_0$ *for the real Lie algebra of* G, *and* $\mathfrak{g}_0^*$ *for its dual. Then the irreducible unitary representations of* G *are in a nat-ural one-to-one correspondence with the set of orbits of* G *on* $\mathfrak{g}_0^*$.

This correspondence has excellent properties with respect to restriction of representations and harmonic analysis.

The *proof* of Theorem 10.1 (in contrast to its statement) introduced no fundamentally new ideas; it is a calculation with the Mackey machine. Our interest lies in the fact that the result is formulated in a way that makes sense when the Mackey machine does not.

To say more, we need some notation. For the time being, G can be an arbitrary Lie group. Of course the action of G on $\mathfrak{g}_0^*$ is the coadjoint action Ad^*: if

λ is a linear functional on g_0, then the linear functional $Ad^*(g)(\lambda)$ is defined by

(10.2) $Ad^*(g)(\lambda)(X) = \lambda(Ad(g^{-1})(X))$.

(If $Ad(g)$ is computed as a matrix in terms of some basis of g_0, then the matrix for $Ad^*(g)$ in terms of the dual basis of g_0^* is the inverse transpose.)

Example 10.3. Suppose G is the three-dimensional Heisenberg group, regarded as real upper triangular three by three matrices with ones on the diagonal. Put

$$g(x,y,z) = \begin{bmatrix} 1 & x & z \\ 0 & 1 & y \\ 0 & 0 & 1 \end{bmatrix}$$

In terms of the basis (e_{12}, e_{23}, e_{13}) of g_0, the adjoint action has matrix

$$Ad(g(x,y,z)) = \begin{bmatrix} 1 & 0 & 0 \\ 0 & 1 & 0 \\ y & -x & 1 \end{bmatrix}$$

The orbits of the adjoint action are therefore the lines

$$ae_{12} + be_{23} + \mathbb{R}e_{13}$$

(for a and b fixed, not both zero); and the points

$$ce_{13}.$$

We get therefore a two-parameter family of lines, and a one-parameter family of points. The coadjoint action has matrix

$$\text{Ad}^*(g(x,y,z)) = \begin{bmatrix} 1 & 0 & -y \\ 0 & 1 & x \\ 0 & 0 & 1 \end{bmatrix}$$

Its orbits are the planes

$$\{\lambda \mid \lambda(e_{13}) = c\},$$

for c a fixed non-zero constant, and the points that take the value 0 on e_{13}. We therefore get planes parametrized by $\mathbb{R}\backslash\{0\}$, and points parametrized by $\mathbb{R}^2$.

The example has two purposes: to show the geometric life that Kirillov's theorem gives to the parametrization of $\hat{G}_u$; and to emphasize that the adjoint and coadjoint actions look quite different in orbit structure.

Suppose again that G is an arbitrary Lie group, and that λ belongs to g_0^*. Write $G(\lambda)$ and $g(\lambda)_0$ for the isotropy group of the coadjoint action at λ, and its Lie algebra. Then

(10.4)(a) $G(\lambda) = \{g \in G \mid \text{Ad}^*(g)(\lambda) = \lambda$

(10.4)(b) $g(\lambda)_0 = \{X \in g_0 \mid \lambda([X,Y]) = 0, \text{ all } Y \in g_0\}.$

An immediate consequence of (10.4)(b) is that λ defines a Lie algebra homomorphism from $g(\lambda)_0$ into $\mathbb{R}$. Recall from

Chapter 1 that we identify the Lie algebra of the circle group with $i\mathbb{R}$.

Definition 10.5. In the setting just described, we say that λ is *integral* if the homomorphism

$$i\lambda: \mathfrak{g}(\lambda)_0 \to \mathrm{Lie}(\mathbb{T})$$

is the differential of a group homomorphism

$$\pi(\lambda)_0: G(\lambda)_0 \to \mathbb{T};$$

that is, of a unitary character of $G(\lambda)_0$.

As Lemma 1.2 might suggest, the point of the integrality assumption is to eliminate the hypothesis that G be simply connected in Theorem 10.1. If G is nilpotent, the kernel of the exponential map is a discrete subgroup of the center of the Lie algebra. We get

COROLLARY 10.6 (to Theorem 10.1). *Suppose G is a connected nilpotent Lie group. Then the irreducible representations of G are in natural one-to-one correspondence with the integral orbits of G on $\mathfrak{g}_0^*$.*

A fairly complete analogue of this result is available for connected type I solvable Lie groups (see [Auslander-Kostant, 1971]. Further generalizations necessarily involve a serious weakening of the conclusions: the complementary

series for $SL(2,\mathbb{R})$ (Theorem 4.23) do not correspond to any

coadjoint orbits.   (There are also problems for compact

groups. The most sophisticated version of the correspon-

dence attaches the trivial representation to each of several

orbits if G is compact but not abelian.) One of the best

results available is that of Duflo (Theorem 0.5); the para-

metrization he gives uses coadjoint orbits and reduces

exactly to that of Kirillov if G is nilpotent. (If G is

reductive, Theorem 0.5 says only that $\hat{G}_u$ is in a natural

one-to-one correspondence with $\hat{G}_u$.)

If we cannot hope to generalize Theorem 10.1 to reduc-

tive groups, we might at least hope for some inspiration.

For example, there ought to be a simple way to attach a

representation to an orbit. Kostant, Duflo, and others have

made great progress on this problem; but it is still nowhere

near a satisfactory resolution. Here is a very brief sketch

of some of their ideas.

Suppose again that G is a Lie group and Λ is an

orbit of G on g_0^*. Write λ for a typical point of Λ,

and use the notation of (10.4). We want to endow the

manifold Λ with a symplectic structure. This means that

we need a non-degenerate symplectic form ω_λ on the tangent

space at each point ω_λ. The forms must vary smoothly with

λ. They will then define a 2-form ω on Λ, and the final requirement for a symplectic structure is that

(10.7)(b) $d\omega = 0.$

The tangent space to a homogeneous space is naturally isomorphic to a quotient of the Lie algebra:

(10.7)(c) $T_\lambda(\Lambda) = \mathfrak{g}_0/\mathfrak{g}(\lambda)_0.$

Define a bilinear form on $\mathfrak{g}_0$ by

(10.7)(d) $\omega_\lambda(X,Y) = \lambda([X,Y]).$

This is obviously skew-symmetric. By (10.4)(b), its radical is precisely $\mathfrak{g}(\lambda)_0$. Conseqently, ω_λ may be regarded as a non-degenerate symplectic form on $T_\lambda(\Lambda)$. That (10.7)(b) is satisfied follows from a short calculation. It is clear from the naturality of the definition that the symplectic structure ω is G-invariant.

One of the things that a coadjoint orbit is, therefore, is a G-space with an invariant symplectic structure. To attach a representation to such things, we should consider how they might arise in connection with a familiar construction. One answer is that the cotangent bundles of a homogeneous space has an invariant symplectic structure . Here is a slight generalization. Suppose that M is a manifold and that $\mathcal{L}$ is a complex line bundle on M. Recall that a *connection* on $\mathcal{L}$ is a map ∇ that assigns to each vector field X on M a first-order differential operator ∇_X on

sections of $\mathcal{L}$. In addition to various linearity properties, ∇ is required to satisfy

(10.8)(a) $\nabla_X(fs) = (X \cdot f)s + f\nabla_X s$,

for f a smooth function and s a smooth section of $\mathcal{L}$.
It follows that the difference of two connections is a map
from vector fields on M to zero-th order differential operators on $\mathcal{L}$; that is,

(10.8)(b) $\nabla_X - \nabla'_X = $ multiplication by g_X;

here the dependence of g on X is C^∞-linear. This means
that g is a section of the complexified cotangent bundle
$T^*(M)_{\mathbb{C}}$.

    If $\mathcal{L}$ is Hermitian, there is a sesquilinear pairing
$\langle \, , \, \rangle$ taking a pair of sections of $\mathcal{L}$ to a function on M.
We can define the adjoint ∇^* of a connection by the re-
quirement

(10.8)(c) $\langle \nabla_X s_1, s_2 \rangle + \langle s_1, \nabla^*_X s_2 \rangle = X \cdot \langle s_1, s_2 \rangle$

for a real vector field X. We say that ∇ is *real* if it
is equal to its adjoint. The difference of two real connections is a section of the real cotangent bundle.

    Arguments like this lead to the following result.

PROPOSITION 10.9 (Urwin; see [Kostant, 1983]). *Suppose* $\mathcal{L}$
is a Hermitian line bundle on an m-*dimensional manifold* M.

Then there is an m-dimensional affine bundle $\mathcal{C} = \mathcal{C}(\mathcal{L})$

over M with the following properties.

a) The vector bundle corresponding to $\mathcal{C}$ is the cotangent bundle.

b) The space of sections of $\mathcal{C}$ is the space of real connections on $\mathcal{L}$.

c) The total space of $\mathcal{C}$ carries a natural symplectic structure.

Assume now that a group G acts on M and $\mathcal{L}$. Then G acts on $\mathcal{C}$, and there is a natural G-equivariant map

$$\mu\colon \mathcal{C} \to \mathfrak{g}_0^*.$$

Assume that there is an open orbit V of G on $\mathcal{C}$. Then the restriction of μ to V is a covering map onto a single coadjoint orbit Λ, respecting the symplectic structures.

(Recall that an affine space A for a vector space V is just a copy of V with the origin forgotten. More formally, A is required to be a principal homogeneous space for V — that is, a homogeneous space for which all isotropy groups are trivial. Affine bundles for vector bundles can now be defined in an obvious way.)

Definition 10.10.   Suppose G is a Lie group and Λ is a

coadjoint orbit. A *real polarization* of Λ is a pair

(M,ℒ) consisting of a homogeneous space and a homogeneous

Hermitian line bundle for G, such that some open orbit of

G on the connection bundle ℭ is a covering of Λ. The

polarization is said to *satisfy the Pukanszky condition* if

G acts transitively on ℭ.

    If a real polarization exists, we say that Λ is *real*

polarizable.

    Polarizations can be described intrinsically.

PROPOSITION 10.11.   *In the setting of Definition 10.10, fix*

λ *in* Λ. *Then real polarizations correspond in a one-to-*

one way to pairs (P,π_p), *subject to the following*

conditions:

    a) P *is a closed subgroup of* G *containing* G(λ)₀

(*cf.* 10.4);

    b) π_p *is a one-dimensional unitary character of* P

with differential iλ|_p; *and*

    c) *The dimension of* G/P *is half the dimension of* Λ.

Notice that only integral orbits can have real polariza-

tions. When a real polarization exists, one can attach to

Λ the representation induced (from P to G) by π_P. Kirillov showed that all coadjoint orbits admit real polarizations in the (simply connected) nilpotent case. This is not so if G is solvable. Auslander and Kostant showed that one could get by with some complex analysis and the following definition.

Definition 10.12. Suppose Λ is a coadjoint orbit for the real Lie group G. Write $\mathfrak{g}$ for the complexified Lie algebra of G. We say that Λ is *algebraically polarizable* if there is a complex subalgebra $\mathfrak{p}$ of $\mathfrak{g}$ such that

a) $\mathfrak{p}$ contains $\mathfrak{g}(\lambda)$;

b) $\lambda([\mathfrak{p},\mathfrak{p}]) = 0$; that is, the restriction of λ to $\mathfrak{p}$ is a one-dimensional character;

c) the dimension of $\mathfrak{g}/\mathfrak{p}$ is half the dimension of $\mathfrak{g}/\mathfrak{g}(\lambda)$; and

d) the sum of $\mathfrak{p}$ and its complex conjugate is a subalgebra $\mathfrak{q}$ of $\mathfrak{g}$.

We call $\mathfrak{p}$ an *algebraic polarization* of Λ at λ. It is called *purely complex* if $\mathfrak{q}$ is all of $\mathfrak{g}$.

Fix a polarization $\mathfrak{p}$. Set

Q_0 = subgroup with Lie algebra $\mathfrak{q}_0$

H_0 = subgroup with Lie algebra $\mathfrak{p} \cap \mathfrak{g}_0$;

the complexified Lie algebra of H is the intersection of

$\mathfrak{p}$ and its complex conjugate. Write Q and H for any subgroups with identity components Q_0 and H_0, such that

   e) H normalizes $\mathfrak{p}$; and

   f) Q contains H.

Here is an outline of how a polarization leads to a representation. By Proposition 1.19, $\mathfrak{p}$ defines a complex structure on $M = Q/H$. If in addition $i\lambda$ is the differential of a character π_H of H, then this character defines a homogeneous holomorphic line bundle $\mathcal{L}$ on M. The way to get a representation in this case is as follows. First, form a representation π_Q of Q on a space of L^2 holomorphic sections of $\mathcal{L}$. Next, induce from Q to G. The main difficulty (which can be avoided if G is solvable) is that $\mathcal{L}$ may have no holomorphic sections. This is essentially the problem discussed in Chapters 5 and 6 when G is reductive. It is by no means completely resolved even in that special case.

A little technical guidance is available from the constructions discussed so far. In the case of a real polarization, we were to induce from the subgroup P. The definition of induction (Definition 3.8) involves twisting the inducing representation by a square root of the modular function for G/P. The examples of Chapters 1, 5, and 6 show that it is reasonable to do that in the case of holomorphic induction as well. In the setting of Definition 10.12, we define the *modular character* of H by

$$(10.13)(a) \qquad \delta_\mathfrak{p}(h) = \det(\mathrm{Ad}(h))\big|_{\mathfrak{g}/\mathfrak{p}}.$$

This is a complex-valued character of H; it is real-valued if the polarization is real, and it has absolute value one if the polarization is purely complex. As in Definition 1.31, $\delta_\mathfrak{p}$ gives rise to a double cover of H on which $\delta_\mathfrak{p}$ has a canonical square root:

$$(10.13)(b) \qquad (\delta_\mathfrak{p})^{\frac{1}{2}} \colon \tilde{H} \to \mathbb{C}^\times.$$

Finally, we arrive at

Definition 10.14. Suppose G is a Lie group, Λ is a coadjoint orbit, and λ is in Λ. A *polarization* of Λ at λ is a quadruple $(\mathfrak{p}, Q, H, \pi)$ with the following properties:

a) $\mathfrak{p}$, Q, and H are as in Definition 10.12; and

b) π_H is an irreducible unitary representation of $\tilde{H}$ (cf. 10.13), such that $d\pi_H$ is a multiple of the restriction of $i\lambda$ to $\mathfrak{h}_0$.

c) π_H takes the value -1 on the non-trivial element ζ of the covering map in (10.13)(b) (cf. Definition 5.7). We say that Λ is *polarizable* if a polarization at λ exists.

To such a set of data, the orbit method can at least attempt to associate a representation (along the lines described after Definition 10.12).

When G is a reductive group, this point of view can lend coherence to the rather oddly assorted constructions of unitary representations presented in Chapters 3 and 6.   It can often suggest important technical improvements; the introduction of metaplectic coverings happened in that way. The ideas around Definition 10.14 provide no more new representations, however.

PROPOSITION 10.15 ([Ozeki-Wakimoto, 1972]). *Suppose* g_0 *is a real reductive Lie algebra and* $\mathfrak{p}$ *is a polarization at* λ *in* g_0^*. *Then* $\mathfrak{p}$ *is a parabolic subalgebra of* g, *and* $\mathfrak{p}$ *has a Levi factor defined over* $\mathbb{R}$.

Because of this proposition, we should concentrate our attention on non-polarizable orbits.   There is no very good systematic theory for attaching representations to non-polarizable coadjoint orbits (but see [Torasso, 1983] or [Guillemin-Sternberg, 1978] for some *ad hoc* successes). What we seek is only guidance about what unipotent representations ought to be, however; so the lack of an actual construction does not make this approach useless.

To begin, we need to know which orbits to look at. The existence of a polarization at λ in the sense of Definition 10.14 does not guarantee that the orbit is integral; it says rather that $i\lambda$ exponentiates to a "metaplectic" character of the two-fold cover of $G(\lambda)_0$ defined by the square root of δ_p (cf. Definition 5.7). We ought therefore to replace the condition of integrality (Definition 10.5) by this one. Unfortunately, the new condition is phrased in terms of a polarization. The next definition cures that problem.

Definition 10.16 (see [Duflo, 1980] or [Duflo, 1982]). Suppose G is a Lie group, and Λ is a coadjoint orbit. Fix a point λ in Λ. Write V_λ for $T_\lambda(\Lambda)$ and ω for the symplectic form on V_λ (cf. (10.7)). Write $\text{Sp}(\omega_\lambda)$ for the group of linear transformations of V_λ preserving ω_λ (the *symplectic group*). The isotropy action then provides a homomorphism

$$\tau_\lambda : G(\lambda) \to \text{Sp}(\omega_\lambda).$$

The symplectic group has a distinguished two-fold covering $\text{Mp}(\omega_\lambda)$ (the *metaplectic group*). If V_λ is zero, it is $\mathbb{Z}/2\mathbb{Z}$; otherwise, it may be characterized as the unique connected two-fold cover. Define $G(\lambda)^\sim$ to be the pullback of the covering $\text{Mp}(\omega_\lambda)$ via τ_λ (Definition 1.31), the *meta-plectic cover* of $G(\lambda)$:

$$1 \to \mathbb{Z}/2\mathbb{Z} \to G(\lambda)^\sim \to G(\lambda) \to 1.$$

Write ζ for the non-trivial element of the kernel of the covering map.

A representation π of $G(\lambda)^\sim$ is called *metaplectic* if $\pi(\zeta) = -I$. It is called *admissible* if it is metaplectic and

$$d\pi(X) = (i\lambda(X)) \cdot I$$

for all X in $g(\lambda)_0$. The orbit Λ is called *admissible* if there exists at least one admissible representation of $G(\lambda)^\sim$.

PROPOSITION 10.17. *Suppose G is a Lie group, and Λ is a polarizable coadjoint orbit (Definition 10.14). Then Λ is admissible. More precisely, the covering (10.13)(b) restricts to the metaplectic covering of $G(\lambda) \cap H$; so a polarization at λ gives rise to an admissible representation of $G(\lambda) \cap H'$.*

The proposition shows that the notion of admissible captures the coverings we want.

We turn now to the study of the structure of coadjoint orbits in the reductive case. For the balance of this section, we return to our usual hypothesis that G is a reductive Lie group. Fix an $Ad(G)$-invariant form $\langle \, , \, \rangle$ on $\mathfrak{g}_0$ as described after (2.1). This induces an isomorphism

(10.18)(a) $\kappa: \mathfrak{g}_0^* \to \mathfrak{g}_0$,

defined by the property that

(10.18)(b) $\langle \kappa(\lambda), Y \rangle = \lambda(Y)$

for all Y in $\mathfrak{g}_0$. We may write

$$X_\lambda = \kappa(\lambda).$$

The map κ is a linear isomorphism, intertwining the coadjoint and adjoint actions:

(10.18)(c) $\kappa(Ad^*(g)(\lambda)) = Ad(g)(\kappa(\lambda))$.

Recall (from [Humphreys, 1972], for example) that every element X of $\mathfrak{g}_0$ has a *Jordan decomposition*

(10.19) $X = X_s + X_n$,

characterized by the properties that $ad(X_s)$ is semisimple (as an automorphism of $\mathfrak{g}_0$); $ad(X_n)$ is nilpotent; X_n belongs to $[\mathfrak{g}, \mathfrak{g}]$; and $[X_s, X_n] = 0$. It follows that X_n acts nilpotently in every finite-dimensional representation and that X_s acts semisimply in every completely reducible finite-dimensional representation. We say that X is *semi-*

simple (respectively nilpotent) if X is equal to X_s (respectively X_n).

We say that an element of $\mathfrak{g}_0^*$ is semisimple (respectively nilpotent) if $\kappa(\lambda)$ is. By (10.18), every element of $\mathfrak{g}_0^*$ has a Jordan decomposition

(10.20) $\lambda = \lambda_s + \lambda_n .$

Here are some of its properties.

PROPOSITION 10.21. Suppose G is a reductive group, and λ belongs to $\mathfrak{g}_0^*$. Write

$$\lambda = \lambda_s + \lambda_n$$

for the Jordan decomposition of λ.

 a) The isotropy group $G(\lambda_s)$ is reductive; in fact it is a Levi subgroup of G (Definition 5.1).

 b) The restriction of λ_n to $\mathfrak{g}(\lambda_s)_0$ is nilpotent. Its isotropy group for the coadjoint action is

$$G(\lambda_s) \cap (\lambda_n) = G(\lambda).$$

 c) The restriction of λ_n to $\mathfrak{g}(\lambda)_0$ is zero.

 d) The G orbit of λ is algebraically polarizable (Definition 10.12) if and only if the $G(\lambda_s)$ orbit of λ_n is algebraically polarizable.

The first three parts of this proposition are routine, and the last is a fairly easy consequence of them. Perhaps the most serious fact one needs to know about the Jordan decomposition in g_0 is that if Y commutes with X, then Y commutes with X_s.

COROLLARY 10.22. *Suppose* G *is reductive and* λ *is a nilpotent element (cf. (10.20) in* g_0^*. *Then the orbit* Λ *of* λ *is integral (Definition 10.5). It is admissible (Definition 10.16) if and only if the metaplectic double cover of* $G(\lambda_s)_0$ *is disconnected.*

Proof. By Proposition 10.21(c), λ is trivial on $g(\lambda)_0$. The trivial character of $G(\lambda)_0$ therefore satisfies the requirement in Definition 10.5. For admissibility, we need a representation π of $G(\lambda)^{\sim}$ with certain properties. It is equivalent to find one π_0 on $G(\lambda)_0^{\sim}$ with these properties. (If we have π_0, then

$$\text{Ind}_{G(\lambda)_0^{\sim}}^{G(\lambda)^{\sim}}(\pi_0)$$

works for π; and if we have π, then its restriction has the right properties for π_0.)

Because λ is trivial on $g(\lambda)_0$, π_0 must be trivial on the identity component of $G(\lambda)_0^{\sim}$. This is compatible with the requirement that $\pi_0(\zeta)$ be -1 exactly when the group is disconnected.   □.

This corollary provides additional evidence that admissibility is a more appropriate condition than integrality. It is fairly well known that certain nilpotent coadjoint orbits ought not to be associated to any unitary representation. An example is the minimal orbit for the symplectic group $Sp(2n,\mathbb{R})$ (with n at least 2). It is associated to the metaplectic representation of the metaplectic double cover $Mp(2n,\mathbb{R})$ but not to any representation of the symplectic group itself. The reason offered by the orbit method is that the orbit is not admissible (except for the covering group).

On the other hand, admissibility alone is still not a sufficient condition to guarantee the existence of a representation attached to the orbit. To see this, take G to be $PSL(2,\mathbb{R})$, and Λ to be a non-zero nilpotent orbit (cf. Example 11.3 below). One can check that Λ is admissible, and even polarizable. (The polarization fails to satisfy the Pukanszky condition, however). There is no representation of $PSL(2,\mathbb{R})$ attached to Λ - the best candidate is a limit of discrete series representation for $SL(2,\mathbb{R})$, and this fails to pass to the quotient $PSL(2,\mathbb{R})$.

The nature of the relationship we want between the orbit method and unipotent representations is this.

"Definition" 10.23. Suppose G is a reductive Lie group. An irreducible representation of G is called *unipotent* if it is a constituent of a representation attached to an admissible nilpotent coadjoint orbit.

This should be compared to Definition 9.19. The quotation marks are needed because we do not know how to define the representation attached to an orbit.

A natural guess is that the set of representations attached to the orbit of λ should be parametrized by the irreducible admissible representations of $G(\lambda)^\sim$ (Definition 10.16). Theorem 1.37 can be interpreted as evidence of this, along with its generalization Theorem 5.12. There are two kinds of problems with the existence of such a parametrization. First, there are the considerations of Chapter 9 for complex groups; these should certainly be compatible with the orbit method. Example 9.22 seeks to attach at least ten different representations to a single orbit; there are only five irreducible admissible representations of $G(\lambda)^\sim$.

There is a much more serious problem, however. Suppose
$M = G/P$ is a homogenous space and $\mathcal{L}$ is a Hermitian line
bundle on M corresponding to a unitary character π_P of
P. Assume that the representation

$$(10.24)(a) \qquad \pi_G = \mathrm{Ind}_P^G(\pi_P)$$

is irreducible. Define the bundle $\mathcal{C}$ of real connections
on $\mathcal{L}$ and the map μ as in Proposition 10.9. Then the
orbit correspondence ought to associate π_G to the image of
μ:

$$(10.24)(b) \qquad \pi_G \leftrightarrow \mu(\mathcal{C}).$$

This image is easily computed from the proof of Proposition
10.9 (which we omitted). The result is

$$(10.24)(c) \qquad \mu(\mathcal{C}) = \{\mathrm{Ad}(g) \cdot \lambda \mid \lambda|_{\mathfrak{p}} = d\pi_P\}.$$

As an example, take G to be $SL(2,\mathbb{R})$, and P the subgroup of upper triangular matrices. Choose π_P to be the
trivial character of P. Then π_G is the spherical principal series representation of $SL(2,\mathbb{R})$ with continuous parameter zero; it is therefore irreducible. The Lie algebra
$\mathfrak{g}_0$, and so also $\mathfrak{g}_0^*$, may be identified with two-by-two
matrices of trace zero. Those restricting to zero on $\mathfrak{p}$
are the strictly lower triangular ones (with zeros on the
diagonal). Consequently, $\mu(\mathcal{C})$ consists of all the conju-
gates of such matrices; and this is the cone of all
nilpotent

elements in $\mathfrak{g}_0$. That cone is the union of three nilpotent orbits, of which two are relatively open.

The conclusion is that representations should correspond not to single nilpotent coadjoint orbits, but to certain closed unions of several orbits. The problem, therefore, is to decide how several nilpotent orbits should be put together to produce something corresponding to a representation. It is easy to imagine that what is involved are the closure relations among the orbits and the singularities of orbit closures. This is a disturbing state of affairs: the orbit closures are only real analytic sets, and one would prefer not to have to say anything clever about their singularities.

There is a way out of this, however. The real nilpotent coadjoint orbits turn out to be in one-to-one correspondence with certain complex algebraic homogeneous spaces; so we can study the problem in the comforting presence of algebraic geometry. This we will do in the next chapter.

Chapter 11

K-MULTIPLICITIES AND UNIPOTENT REPRESENTATIONS

Since the orbit method does not yet provide a construction of representations attached to nilpotent orbits, we
need a less direct way to guess what those representations
ought to be. Our main approach will be through the restriction of the representation to the maximal compact subgroup
K: we will try to read off from the orbit what this restriction ought to be.

Fix a reductive group G (Definition 0.6), and choose K
and θ as at the beginning of Chapter 2. (We will eventually need G to be in Harish-Chandra's class, but this assumption can be omitted at the beginning.) Put

(11.1)(a) $K_{\mathbb{C}}$ = complexification of K;

this is a complex reductive algebraic group (possibly disconnected) which acts algebraically on any locally finite representation of K. In particular,

258

(11.1)(b) $K_{\mathbb{C}}$ acts algebraically on any $(\mathfrak{g},K)$-module.

Of course $K_{\mathbb{C}}$ need not be connected. Fix a complex connected reductive algebraic group $G_{\mathbb{C}}$ with Lie algebra $\mathfrak{g}$.
This group need not contain G, and the notation is there-
fore misleading; but we will make very little use of it in
any case. Put

(11.1)(c) $\mathfrak{s}_0 = -1$ eigenspace of θ in $\mathfrak{g}_0$;

dropping the zero will denote complexification as usual. We
often identify $\mathfrak{s}$ with $\mathfrak{g}/\mathfrak{k}$. For example, this allows us
to regard linear functionals on $\mathfrak{s}$ as functionals on $\mathfrak{g}$
vanishing on $\mathfrak{k}$, and gives

(11.1)(d) $\mathfrak{s}^* \subset \mathfrak{g}^*.$

    Define

(11.2)(a) $\mathscr{N}_{\mathbb{C}} =$ cone of nilpotent elements in $\mathfrak{g}^*$

(defined before (10.20)). The group $G_{\mathbb{C}}$ acts on $\mathscr{N}_{\mathbb{C}}$; its
orbits there were considered at the end of Chapter 9. Put

(11.2)(b) $\mathscr{N}_{\mathbb{R}} = \mathscr{N}_{\mathbb{C}} \cap \mathfrak{g}_0^*,$

the cone of nilpotent elements in $\mathfrak{g}_0^*$. The group G acts
on $\mathscr{N}_{\mathbb{R}}$; its orbits are the ones in Definition 10.23.
Finally, put

(11.2)(c) $\mathscr{N}_{\theta} = \mathscr{N}_{\mathbb{C}} \cap \mathfrak{s},$

the cone of nilpotent elements in $\mathfrak{s}^*$. The group $K_{\mathbb{C}}$ acts
on $\mathscr{N}_{\theta}$; its orbits are going to be the objects of our atten-

tion now. The advantage of $\mathcal{N}_\theta$ over $\mathcal{N}_{\mathbb{R}}$ is that the former is an algebraic variety, and the action of the group $K_{\mathbb{C}}$
is algebraic.

Example 11.3. Define G^1 to be $SL(2,\mathbb{R})$. The Lie algebra
$\mathfrak{g}_0^1$ of G^1 consists of two by two real matrices of trace
zero; its complexification consists of complex matrices of
trace zero. We can choose

$$(\theta^1)g = {}^tg^{-1} \qquad (g \in G)$$

$$(\theta^1)X = -{}^tX \qquad (X \in \mathfrak{g})$$

$$K^1 = SO(2)$$

$$(K^1)_{\mathbb{C}} = SO(2,\mathbb{C})$$

$$G_{\mathbb{C}}^1 = SL(2,\mathbb{C})$$

$$\mathfrak{s}^1 = \text{symmetric matrices in } \mathfrak{g}$$

A two by two matrix of trace zero is nilpotent if and
only if it has determinant zero. Consequently

$$\mathcal{N}_{\mathbb{C}}^1 = \left\{ \begin{bmatrix} a & b \\ c & -a \end{bmatrix} \,\middle|\, a^2 + bc = 0 \right\}$$

Here a, b, and c are complex. The group $G_{\mathbb{C}}$ has exactly
two orbits on $\mathcal{N}_{\mathbb{C}}^1$: the point zero, and everything else. As
a representative of the non-zero orbit, we can choose

$$X_{\mathbb{C}}^1 = \begin{bmatrix} 0 & 1 \\ 0 & 0 \end{bmatrix}$$

The real nilpotent cone looks just like $\mathcal{N}_{\mathbb{C}}^1$, except that the entries a, b, and c must now be real numbers. The equation $a^2 + bc = 0$ has no non-zero solutions with b equal to c; so if we set

$[\mathcal{N}_{\mathbb{R}}^1]^+$ = real matrices as above with b > c,

and define $[\mathcal{N}_{\mathbb{R}}^1]^-$ similarly, then

$$\mathcal{N}_{\mathbb{R}}^1 = [\mathcal{N}_{\mathbb{R}}^1]^+ \cup \{0\} \cup [\mathcal{N}_{\mathbb{R}}^1]^-.$$

Because G^1 is connected, it must respect this decomposition; and it is easy to check that these are precisely the three orbits of G. As representatives of the two large orbits, we can choose

$$X_{\mathbb{R}}^1 = \begin{bmatrix} 0 & 1 \\ 0 & 0 \end{bmatrix}$$

$$Y_{\mathbb{R}}^1 = \begin{bmatrix} 0 & 0 \\ 1 & 0 \end{bmatrix}$$

The nilpotent cone in $\mathfrak{s}^1$ consists of trace zero complex symmetric matrices of determinant zero. Set

$$X_{\theta}^1 = \begin{bmatrix} 1 & i \\ i & -1 \end{bmatrix}$$

$$Y_\theta^1 = \begin{bmatrix} 1 & -i \\ -i & -1 \end{bmatrix}$$

Define

$$[\mathcal{N}_\theta^1]^+ = \mathbb{C}^\times \cdot X_\theta^1,$$

and define $[\mathcal{N}_\theta^1]^+$ similarly. Then it is easy to check that these two sets and zero are the orbits of $K_\mathbb{C}^1$ on $\mathcal{N}_\theta^1$.

This example is fundamental, particularly in light of the Jacobson–Morozov theorem and its descendants.

THEOREM 11.4 (Jacobson–Morozov; see [Kostant, 1959]). *Suppose* $G_\mathbb{C}$ *is a complex reductive group with Lie algebra* $\mathfrak{g}$. *Then the finite set of nilpotent orbits of* $G_\mathbb{C}$ *on* $\mathfrak{g}$ *is in one-to-one correspondence with the set of* $G_\mathbb{C}$*-conjugacy classes of Lie algebra homomorphisms*

$$\psi_\mathbb{C} \colon \mathfrak{g}^1 \to \mathfrak{g};$$

here $\mathfrak{g}^1$ *is* $\mathfrak{sl}(2,\mathbb{C})$, *as in Example 11.3. The correspondence sends the homomorphism* $\psi_\mathbb{C}$ *to the nilpotent element* $\psi_\mathbb{C}(X_\mathbb{C}^1)$ *(defined in Example 11.3).*

This result is true over $\mathbb{R}$ as well. We prefer to phrase

that fact in a slightly roundabout way, to emphasize the
analogy with Theorem 11.6 below.

THEOREM 11.5 (Jacobson-Morozov; see [Kostant, 1959].) *Sup-*
pose G *is a real reductive group with Lie algebra* $\mathfrak{g}_0$.
Use the notation of Example 11.3. Then the finite set of
nilpotent orbits of G *on* $\mathfrak{g}_0$ *is in one-to-one correspon-*
dence with the set of G-*conjugacy classes of Lie algebra*
homomorphisms

$$\psi_{\mathbb{C}}: \mathfrak{g}^1 \to \mathfrak{g}$$

that respect the complex conjugations σ^1 *and* σ *on* $\mathfrak{g}^1$
and $\mathfrak{g}$:

$$\psi_{\mathbb{R}}(\sigma^1 A) = \sigma(\psi_{\mathbb{R}}(A)).$$

The correspondence sends $\psi_{\mathbb{R}}$ *to* $\psi_{\mathbb{R}}(X_{\mathbb{R}}^1)$.

   If X *is a nilpotent element of* $\mathfrak{g}_0$, *then the real*
dimension of G·X *is equal to the complex dimension of*
$G_{\mathbb{C}}$·X.

   The corresponding result for nilpotents in $\mathfrak{s}$ is due
to Kostant and Rallis.

THEOREM 11.6 ([Kostant-Rallis, 1971]). *Suppose* G *is a*
real reductive Lie group; use the notation of (11.1),
(11.2), and Example 11.3. Then the finite set of nilpotent

orbits of $K_{\mathbb{C}}$ on $\mathfrak{s}$ is in one-to-one correspondence with the set of $K_{\mathbb{C}}$-conjugacy classes of Lie algebra homomorphisms

$$\psi_\theta \colon \mathfrak{g}^1 \to \mathfrak{g}$$

intertwining the actions of θ^1 and θ:

$$\psi_\theta(\theta^1 A) = \theta(\psi_\theta(A)).$$

The correspondence sends ψ_θ to $\psi_\theta(X^1_\theta)$.

Fix a nilpotent element X in $\mathfrak{s}$. Then the complex dimension of the $K_{\mathbb{C}}$ orbit $K_{\mathbb{C}} \cdot X$ is half the dimension of the $G_{\mathbb{C}}$ orbit $G_{\mathbb{C}} \cdot X$.

The next result provides the formal relationship we want between $N_{\mathbb{R}}$ and N_θ.

THEOREM 11.7 (Sekiguchi). Suppose G is a reductive algebraic group; use the notation of (11.1), (11.2), and Example (11.3).

Suppose first that $\psi_{\mathbb{R}}$ is a homomorphism of $\mathfrak{g}^1$ into $\mathfrak{g}$, respecting the complex conjugations. Then there is a G conjugate $\psi_{\mathbb{R}\theta}$ of $\psi_{\mathbb{R}}$ that also intertwines the actions of θ^1 and θ. This conjugate is unique up to conjugation by K (the θ-fixed elements of G).

On the other hand, suppose that $\psi_{\mathbb{R}}$ is a homomorphism of $\mathfrak{g}^1$ into $\mathfrak{g}$ intertwining θ^1 and θ. Then there is a

$K_{\mathbb{C}}$ conjugate $\psi_{\theta\mathbb{R}}$ of ψ_{θ} that also respects the complex conjugations. This conjugate is unique up to conjugation by K (the elements of $K_{\mathbb{C}}$ fixed by complex conjugation).

COROLLARY 11.8. Suppose G is a real reductive group; use the notation of (11.1) and (11.2). Then there is a natural bijection between the orbits of G on $\mathcal{N}_{\mathbb{R}}$ and the orbits of $K_{\mathbb{C}}$ on $\mathcal{N}_{\theta}$. If $\Lambda_{\mathbb{R}}$ corresponds to Λ_{θ} in this bijection, then

$$G_{\mathbb{C}}\Lambda_{\mathbb{R}} = G_{\mathbb{C}}\Lambda_{\theta}.$$

It is not a trivial matter to write down the map of Corollary 11.8 in either direction; one really has to pass through the special $\mathfrak{sl}(2)$ homomorphisms. It seems very likely that the bijection preserves the closure relation on the orbits, but I do not know how to prove such an assertion.

Definition 11.9 (cf. [Vogan, 1978]). In the setting (11.1), suppose X is a finitely generated Harish-Chandra module. Recall from (11.1)(b) that $K_{\mathbb{C}}$ acts on X. A *good filtration* of X is a (possibly infinite) increasing filtration

(a) $X_0 \subset X_1 \subset \ldots \subset X$

of X, satisfying the conditions below. Write U_n for the

nth level of the standard filtration of $U(\mathfrak{g})$. By the

Poincaré-Birkhoff-Witt theorem, the associated graded ring

$gr(U(\mathfrak{g}))$ is naturally isomorphic to $S(\mathfrak{g})$. The conditions

are as follows:

(i) X_m is finite-dimensional and $K_{\mathbb{C}}$-invariant.

(ii) The union of all the X_m is all of X.

(iii) The filtrations of X and $U(\mathfrak{g})$ are compatible:

$$U_n X_m \subseteq X_{n+m}.$$

(iv) The associated graded $S(\mathfrak{g})$-module $gr(X)$ (which

makes sense by (iii)) is finitely generated.

Good filtrations certainly exist. To see this, choose

any finite-dimensional generating subspace S of X. By

Definition 1.26(b), S is contained in a finite-dimensional

$K_{\mathbb{C}}$-invariant subspace X_0. Set

(b) $X_m = U_m X_0.$

This is easily seen to be a good filtration.

For any good filtration, the associated graded module

$gr(X)$ is a finitely generated $S(\mathfrak{g})$-module, equipped with a

compatible algebraic action of $K_{\mathbb{C}}$. (Compatibility is de-

fined in analogy with Definition 1.26(c).) We will express

this by calling $gr(X)$ an $(S(\mathfrak{g}), K_{\mathbb{C}})$-module. The module

$gr(X)$ depends on the choice of good filtration, but only a

little; for example, its class in the Grothendieck group of

finitely generated $(S(\mathfrak{g}), K_{\mathbb{C}})$ modules is well-defined.
Accordingly the support of gr(X)

$$\text{Supp}(\text{gr}(X)) \subset \text{Spec } S(\mathfrak{g}) = \mathfrak{g}^*$$

depends only on X, and not on the filtration. (Recall that
the support of a module M over a commutative ring consists
of those prime ideals at which the localization of M is
not zero. For finitely generated modules over Noetherian
rings, this is the same as the associated variety of the
annihilator of M.) We write

(c)   Ass(X) = Supp(gr(X)),

and call this the *associated variety of* X.

PROPOSITION 11.10 (see [Vogan, 1978]). *Suppose* X *is a*
finitely generated Harish-Chandra module. Then Ass(X)
(Definition 11.9) is a $K_{\mathbb{C}}$-*invariant closed cone in* $\mathfrak{s}^*$
(cf. (11.1)). If X *has finite composition series, then*
Ass(X) *is contained in the nilpotent cone* $\mathcal{N}_{\theta}$ *(cf.*
(11.2)).

A deep theorem of Gabber says that if X is irreducible,
then Ass(X) is equidimensional. (It need not be irreducible.) We will make no use of this result.

We have now attached to any Harish–Chandra module X
of finite length a closed union Ass(X) of nilpotent $K_{\mathbb{C}}$
orbits on $\mathfrak{s}$. Because of the bijection of Corollary 11.8,
we also get a union of real nilpotent orbits. Here is how
this correspondence should be related to the method of coadjoint orbits.

(False) "Conjecture" 11.11 (see [Barbasch–Vogan, 1980]).
Suppose $\Lambda_{\mathbb{R}}$ is a finite union of orbits in $\mathfrak{g}_0^*$. The
associated cone Ass($\Lambda_{\mathbb{R}}$) is the set of all limits of con-
vergent sequences
$$(s_i \lambda_i) \quad (s_i \in \mathbb{R}^+,\ \lambda_i \in \mathfrak{g}_0^*,\ s_i \to 0).$$
It is a G–invariant cone contained in $\mathcal{N}_{\mathbb{R}}$, and therefore a
finite union of nilpotent G orbits. If $\Lambda_{\mathbb{R}}$ is closed and
nilpotent, then
$$\mathrm{Ass}(\Lambda_{\mathbb{R}}) = \Lambda_{\mathbb{R}}.$$
Write $\mathrm{Ass}(\Lambda_{\mathbb{R}})_\theta$ for the corresponding union of $K_{\mathbb{C}}$
orbits on $\mathcal{N}_\theta$ (Corollary 11.8). Suppose X is a unitary
$(\mathfrak{g},K)$–module associated to $\Lambda_{\mathbb{R}}$ by the orbit method. Then
$$\mathrm{Ass}(X) = \mathrm{Ass}(\Lambda_{\mathbb{R}})_\theta.$$

The reason this is false is that the holomorphic part of the
orbit method (in our case, the cohomological parabolic induc-

tion functors) can attach the zero representation to a non-
empty orbit. That problem should not arise for nilpotent orbits; so one can drop the "false" part if $\Lambda_{\mathbb{R}}$ is contained
in $\mathcal{N}_{\mathbb{R}}$. The reason that for the quotation marks is that
there is no definition in general for the representation
"associated" to an orbit or collection of orbits. From now
on, we will interpret statements like this not as conjectures, but as partial specifications for a future orbit
method.

In addition to Ass(X), we have the $(S(\mathfrak{g}), K_{\mathbb{C}})$-module
gr(X), supported on Ass(X). This is not quite so well de-
fined as its support, since it depends on the choice of good
filtration. The main observation is that this module charac-
terizes the restriction of our original X to K:

Observation 11.12. Suppose X is any $(\mathfrak{g}, K)$-module of
finite length. Then X is isomorphic as a representation
of K (or $K_{\mathbb{C}}$) to the $(S(\mathfrak{g}), K_{\mathbb{C}})$-module gr(X).

This is a trivial consequence of the complete reducibility
of finite-dimensional representations of compact groups. In
conjunction with Conjecture 11.11, it leads to the following
constraint on the orbit method.

Requirement 11.13. Suppose $\Lambda_{\mathbb{R}}$ is a union of G orbits on $\mathcal{N}_{\mathbb{R}}$. Write Λ_{θ} for the corresponding union of $K_{\mathbb{C}}$ orbits on $\mathcal{N}_{\theta}$. If X is a $(\mathfrak{g},K)$-module associated to $\Lambda_{\mathbb{R}}$ by the orbit method, then there must be an $(S(\mathfrak{g}),K_{\mathbb{C}})$-module M, such that

  i) $\mathrm{Supp}(M) = \Lambda_{\theta}$; and

  ii) $M \cong X$ as a representation of K.

So far this is a fairly weak requirement. We intend to sharpen it by sharpening the requirements on M.

To see how this ought to be done, recall that we already know that representations ought to correspond only to admissible orbits. We will therefore investigate which orbits on $\mathcal{N}_{\theta}$ correspond to admissible orbits on $\mathcal{N}_{\mathbb{R}}$; then try to find a kind of module that can be supported only on these orbits.

So let $\lambda_{\mathbb{R}}$ be an element of the G orbit $\Lambda_{\mathbb{R}}$, corresponding to $X_{\mathbb{R}}$ in $\mathfrak{g}_0$ (cf. 10.18). After conjugating by an element of G, we can and do assume that there is a map

(11.14)(a) $\qquad\qquad \psi \colon \mathfrak{sl}(2) \to \mathfrak{g}$,

respecting both the Cartan involutions and the complex conjugations, such that

(11.14)(b) $\qquad\qquad \psi(X_{\mathbb{R}}^1) = X_{\mathbb{R}}$

(cf. Example 11.3). Define

(11.14)(c) $X_\theta = \psi(X_\theta^1)$

(11.14)(d) λ_θ = corresponding element of $\mathfrak{s}^*$.

Then λ_θ is a representative of the $K_\mathbb{C}$ orbit Λ_θ corresponding to $\Lambda_\mathbb{R}$.

To simplify the notation (at the small expense of confusing the reader), we will write $H_\mathbb{R}$ for the isotropy group called $G(\lambda_\mathbb{R})$ in (10.4):

(11.15)(a) $H_\mathbb{R} = \{h \in G \mid Ad^*(g)(\lambda_\mathbb{R}) = \lambda_\mathbb{R}\}$.

Similarly, write

(11.15)(b) $H_\theta = \{h \in K_\mathbb{C} \mid Ad^*(k)(\lambda_\theta) = \lambda_\theta\}$.

Each of these groups contains

(11.15)(c) $H = \{h \in K \mid Ad(k)(\psi) = \psi\}$;

it can be shown that H is a maximal compact subgroup of each.

The complex group H_θ acts algebraically on the algebraic (or, equivalently, holomorphic) cotangent space at λ_θ to the complex variety Λ_θ. The determinant of this action is an algebraic character of H_θ, which we call δ_θ. Formally,

(11.16)(a) $\delta_\theta(h) = \det(Ad^*(h) \text{ on } (\mathfrak{k}/\mathfrak{h}_\theta)^*)$.

Let

(11.16)(b) $H_\theta^\sim$ = two-fold cover of H_θ attached to the

square root of δ_θ.

Definition 10.16 describes the two-fold metaplectic cover $\widetilde{H_{\mathbb{R}}}$ of $H_{\mathbb{R}}$.

The following result will appear in the M.I.T. doctoral dissertation of J. Schwartz.

PROPOSITION 11.17. *With notation as above, the restrictions to H of the coverings* $\widetilde{H_{\mathbb{R}}}$ *and* $\widetilde{H_{\theta}}$ *are naturally isomorphic. Write* $\widetilde{H}$ *for this common restriction. Then the following sets are in natural one-to-one correspondence:*

    i) admissible irreducible representations of $\widetilde{H_{\mathbb{R}}}$;

    ii) admissible irreducible representations of $\widetilde{H}$;

    iii) admissible irreducible representations of $\widetilde{H_{\theta}}$.

In each case, admissible means trivial on the identity component and non-trivial on the kernel of the covering map.

It is easy to describe the genuine representations of a covering attached to the square root of a character. We can therefore deduce

COROLLARY 11.18. *In the setting (11.14)-(11.16), the set of irreducible admissible representations of* $H_{\mathbb{R}}$ *is in one-to-one correspondence with the set of irreducible admissible* algebraic representations ϕ of H_θ satisfying

$$d\phi = \tfrac{1}{2}d\delta_\theta.$$

To understand the significance of this result, we need
an algebraic analogue of Proposition 3.2.

PROPOSITION 11.19. Suppose G is a complex algebraic group
and H is a closed subgroup. Write $\mathcal{O}_{G/H}$ for the sheaf of
regular functions on the algebraic variety G/H. Then the
following three sets are in natural one-to-one correspondence.

   i) Coherent $\mathcal{O}_{G/H}$ modules with an algebraic action
of G compatible with the action on G/H (briefly,
$(\mathcal{O}_{G/H},G)$-modules).

   ii) Homogeneous algebraic vector bundles on G/H.

   iii) Algebraic representations of H.

Suppose G/H is an affine variety, and write $R(G/H)$ for
the ring of regular functions on G/H (the global sections
of $\mathcal{O}_{G/H}$). Then these sets are also in one-to-one correspondence with

   iv) finitely generated $(R(G/H),G)$-modules (defined as
in Definition 1.26(c)).

The proof is formally identical to that of Proposition 3.2. The last part is the usual correspondence between modules and sheaves for affine algebraic varieties.

COROLLARY 11.20. *In the setting* (11.14)-(11.16), *write* $\mathcal{D}_\theta$ *for the top exterior power of the cotangent bundle on* Λ_θ. *Then* $\Lambda_\mathbb{R}$ *is admissible if and only if there is a* $K_\mathbb{C}$-*equivariant vector bundle* $\mathscr{V}$ *on* Λ_θ, *satisfying one of the following equivalent conditions:*

   i) $\mathscr{V} \otimes \mathscr{V}$ *is isomorphic (as an equivariant vector bundle) to a sum of copies of* $\mathcal{D}_\theta$, *after both bundles have been lifted to some* $(K_\mathbb{C}$-*equivariant*) *finite cover of* Λ_θ.

   ii) There is an equivariant locally constant sheaf of vector spaces $\mathscr{F}$ *on* Λ_θ *such that*
$$\mathscr{V} \otimes \mathscr{V} \cong \mathcal{D}_\theta \otimes \mathscr{F}.$$
Here we interpret a locally constant sheaf as a vector bundle with a flat connection.

The problems discussed around Definition 10.23 suggest that we need to extend conditions like those in Corollary 11.20 to the closure of an orbit, or even to a closed union of several orbits. We know that (coherent sheaves of) modules are a reasonable generalization of vector bundles to this setting (cf. Requirement 11.13). It is less clear what ought to replace the top exterior power of the cotangent bundle. On the basis of what happens in SL(2), Deligne has suggested that the right object is the *dualizing complex*. Suppose V is a (possibly reducible) algebraic variety. Put

(11.21)(a) O_V = sheaf of regular functions on V;

when no confusion is possible, we call this O. Then there is a coherent sheaf of O-modules

(11.21)(b) $\mathscr{D}_V$ = dualizing sheaf.

Restricted to the top-dimensional part of the smooth locus of V, $\mathscr{D}_V$ is isomorphic to the (sheaf of germs of sections of) the top exterior power of the cotangent bundle. Even more is true: there is a complex $\omega^{\cdot}$ of sheaves (or rather an object in a derived category) of which $\mathscr{D}_V$ is the top degree cohomology. The suggestions below will be stated in terms of $\mathscr{D}$ for simplicity; they can be reformulated using $\omega^{\cdot}$ without difficulty. Since the suggestions are so imprecise in any case, this refinement is of little importance. Define

(11.21)(c)   R(V) = ring of global sections of O_V.

If V is affine, we can as usual identify the sheaf $\mathscr{D}$ with the R(V)-module

(11.21)(d)   D(V) = global sections of $\mathscr{D}_V$.

At this point I do not know how to proceed. Here is one possibility.

Lack of Definition 11.22. Suppose V is an algebraic variety; use the notation of 11.23. There should be a notion of *self-dual* for a coherent $\mathcal{O}$-module $\mathcal{M}$. I do not know exactly what it is, but it should probably satisfy the following constraints:

a) If $\mathcal{M} \otimes_{\mathcal{O}} \mathcal{M}$ is isomorphic to a multiple of the dualizing sheaf $\mathcal{D}$, then M is self-dual.

b) If $\mathcal{M}$ is self-dual, then there is a locally constant sheaf $\mathcal{F}$ of vector spaces on V, such that

$$\mathcal{M} \otimes_{\mathcal{O}} \mathcal{M} \cong \mathcal{F} \otimes_{\mathcal{O}} \mathcal{D}.$$

Here now is a strengthening of Requirement 11.13.

Requirement 11.23. Suppose $\Lambda_{\mathbb{R}}$ is a union of G orbits on $\mathcal{N}_{\mathbb{R}}$. Write Λ_{θ} for the corresponding union of $K_{\mathbb{C}}$ orbits on $\mathcal{N}_{\theta}$. If X is a $(\mathfrak{g},K)$-module associated to $\Lambda_{\mathbb{R}}$ by the orbit method, then the following conditions must be satisfied:

a) Λ_{θ} is closed and equidimensional.

b)   There is a finitely generated $(S(\mathfrak{g}), K_{\mathbb{C}})$-module M
satisfying

$$M \cong X \text{ as a representation of } K$$
$$\text{Ann } M = \text{ideal of } \mathcal{O}_{\theta}$$
$$M \text{ is self-dual (in a } K_{\mathbb{C}}\text{-equivariant way)}$$
$$\text{as a module on } \Lambda_{\theta}.$$

Because of the remark before Definition 11.22, this requirement should not be taken too seriously.

In the same spirit, here is a tentative outline of what
is wanted from the orbit method for nilpotent orbits.

Conjecture 11.24.   Suppose G is a reductive group in
Harish-Chandra's class. Use the notation (11.1) and (11.2);
assume that $G_{\mathbb{C}}$ is chosen so that the adjoint representation of G factors through $G_{\mathbb{C}}$. Fix

i)   the closure $\Lambda_{\mathbb{C}}$ of a single $G_{\mathbb{C}}$ orbit on $\mathcal{N}_{\mathbb{C}}$;

ii) an irreducible, normal, affine algebraic variety
$V_{\mathbb{C}}$ equipped with

1) an algebraic action of $G_{\mathbb{C}}$, and

2) a finite, $G_{\mathbb{C}}$-equivariant morphism π from $V_{\mathbb{C}}$
onto $\Lambda_{\mathbb{C}}$;

iii)   a closed, $K_{\mathbb{C}}$-invariant equidimensional subvariety

$$V_\theta \subset \pi^{-1}(\Lambda_{\mathbb{C}} \cap s^*),$$

of half the dimension of $V_{\mathbb{C}}$; and

iv)   a self-dual (Definition 11.22) $(R(V_\theta),K_{\mathbb{C}})$ module M.

Then there is (conjecturally) a unitary $(\mathfrak{g},K)$-module

$$X = X(V_\theta,M),$$

such that

a) $X \cong M$ as a representation of K.

More precisely, write A for the completely prime primitive algebra (conjecturally) associated to $V_{\mathbb{C}}$ by Conjecture 9.17. We should have

b)   X is an $(A,K_{\mathbb{C}})$-module.

The algebra A should admit a conjugate linear anti-automorphism

$$a \to a^h$$

extending the one on $U(\mathfrak{g})$ (Definition 9.9). Write

$$k \to (k^h)^{-1}$$

for the complex conjugation on $K_{\mathbb{C}}$ with fixed point set K. Then X should be endowed with a positive definite sesquilinear form $\langle \, , \, \rangle$ satisfying

c) $\langle av,w \rangle = \langle v,a^h w \rangle$

for v and w in X, and a in A or $K_{\mathbb{C}}$.

One can easily impose additional conditions, saying for example that some associated graded module gr(X) should be isomorphic to M, or to some gr(M). Probably X should be irreducible (as an $(A, K_{\mathbb{C}})$-module) whenever M is (in some self-dual-module-theoretic sense). Almost certainly $R(V_{\mathbb{C}})$ has a Poisson bracket structure; at any rate the largest $G_{\mathbb{C}}$-orbit in $V_{\mathbb{C}}$, whose complement has codimension at least two, is a symplectic manifold. Hypothesis (iii) of the conjecture makes V_θ a Lagrangian subvariety, and this is certainly a good thing.

"Definition" 11.25. Suppose G is a reductive group in Harish-Chandra's class. A representation of G is called unipotent if it is a sum of constituents of various representations $X(V_\theta, M)$ as in Conjecture 11.24.

Notice that a necessary condition for X to be unipotent is that the annihilator of X be a self-adjoint unipotent primitive ideal (cf. Conjecture 9.19). Unfortunately, there are non-unitary representations that satisfy this condition. The non-spherical principal series for $SL(2, \mathbb{R})$ having continous parameter $\frac{1}{2}\rho$ is an example; its annihilator is the

same as that of the metaplectic representation of $SL(2,\mathbb{R})^{\sim}$.
It is therefore not satisfactory to use Definition 9.20 for
real groups.

Example 11.26. Write ω for the usual symplectic form on
$\mathbb{R}^{2n}$. To describe it, write

$$\mathbb{R}^{2n} = \{(x,y) \mid x,y \in \mathbb{R}^n\}$$

$$\omega((x,y),(z,w)) = \langle x,w \rangle - \langle y,z \rangle,$$

with $\langle\ ,\ \rangle$ the usual inner product on $\mathbb{R}^n$. All of this
works over $\mathbb{C}$ as well; we have to use the symmetric (as
opposed to Hermitian) form $\langle\ ,\ \rangle$. Put

$$G = Sp(2n,\mathbb{R})$$

$$G_{\mathbb{C}} = Sp(2n,\mathbb{C})$$

The Cartan involution θ may be taken to be

$$\theta X = -{}^t X \qquad (X \in \mathfrak{g})$$

Then the maximal compact subgroup K of G is isomorphic
to the unitary group $U(n)$, and $K_{\mathbb{C}}$ is $GL(n,\mathbb{C})$.

Put

$$V_{\mathbb{C}} = \mathbb{C}^{2n},$$

with the standard action of $G_{\mathbb{C}}$. Define a map π from $V_{\mathbb{C}}$
to $\mathfrak{g}$ (which is a space of linear transformations of $\mathbb{C}^{2n}$)
by

$$\pi(v)w = \omega(v,w)v.$$

This is equivariant and finite. The corresponding algebra

A (Conjecture 9.17) is the Weyl algebra of polynomial coefficient differential operators on $\mathbb{C}^n$. We can identify the generators of A with a basis of $\mathbb{C}^{2n}$ by

$$x \rightarrow (0,e_i)$$
$$\partial/\partial x_j \rightarrow (e_j,0)$$

Write $A_{(1)}$ for the linear span of these generators, and write η for this identification of $A_{(1)}$ with $\mathbb{C}^{2n}$. The defining relations for A (the canonical commutation relations) may now be written

$$[r,s] = \omega(\eta(r),\eta(s)) \qquad (r,s \in A_{(1)}).$$

The group $G_{\mathbb{C}}$ therefore acts by automorphisms on A, by

$$g \cdot r = \eta^{-1}(g \cdot \eta(r)).$$

We want to define a Lie algebra homomorphism ϕ from $\mathfrak{g}$ into A. First, we give an isomorphism ψ from $S^2(\mathbb{C}^{2n})$ (regarded as symmetric 2-tensors) onto $\mathfrak{g}$. This is

$$\psi(u \otimes v + v \otimes u)w = \omega(u,w)v + \omega(v,w)u.$$

Next, we write μ for the multiplication map from $S^2(A_{(1)})$ into A. Finally, we set

$$\phi(X) = \mu(\psi^{-1}(X)) \qquad (X \in \mathfrak{g}).$$

It is easy to see that this respects the action of $G_{\mathbb{C}}$; we leave to the reader the task of verifying that ϕ is a Lie algebra homomorphism.

It is easy to verify that the set of elements of $V_{\mathbb{C}}$ mapping into s^* under π is

$$\{(x, \pm ix) \mid x \in \mathbb{C}^n\}.$$

The group $K_{\mathbb{C}}$ (which is $GL(n, \mathbb{C})$) acts on this set by acting on x in the usual way. Set

$$V_\theta = \{(x, ix)\}$$
$$\cong \mathbb{C}^n.$$

Because this is smooth, the dualizing module D is the module of polynomial-coefficient top-degree holomorphic differential forms on V_θ. If we write

$$d\xi = dx_1 \wedge \cdots \wedge dx_n,$$

then $K_{\mathbb{C}}$ acts on D by

$$g \cdot f(x) d\xi = h(x) d\xi,$$

with

$$h(x) = \det(g) h(g^{-1} \cdot x).$$

It is fairly easy to deduce that there are no self-dual $(R(V_\theta), K_{\mathbb{C}})$-modules. (If we forget the $K_{\mathbb{C}}$ action, the dualizing module is isomorphic to the free module M on one generator; and it follows that M itself is self-dual. The isomorphism cannot be made equivariant, however.)

Suppose now that G is replaced by its two-fold cover $\tilde{G} = Mp(2n, \mathbb{R})$. This replaces $K_{\mathbb{C}}$ by the cover attached to the square root of the determinant character. Now let M be the free $R(V_\theta)$-module on one generator ζ. We can make M into an $(R(V_\theta), (K_{\mathbb{R}})^{\tilde{}})$-module, by making $\tilde{K}_{\mathbb{C}}$ act by

$$g \cdot f(x) \zeta = h(x) d\zeta,$$

with

$$h(x) = \det^{\frac{1}{2}}(g)h(g^{-1} \cdot x).$$

It is easy to see that $M \otimes_R M$ is isomorphic to D. We therefore have the data for Conjecture 11.24, and can look for a $(g, \tilde{K})$-module X. The Harish-Chandra module of the metaplectic representation fulfills the requirements (a)–(c); we will omit the verification of this fact. Explicitly, X can be regarded as a space of functions on $\mathbb{R}^n$:

$$X = \{p(x)e^{-\langle x \cdot x \rangle} \mid p \text{ is a polynomial}\}.$$

X is irreducible as a module for the Weyl algebra A but splits as a sum of two pieces (the even and odd functions) as a Harish-Chandra module.

Chapter 12

ON THE DEFINITION OF UNIPOTENT REPRESENTATIONS

Having consulted our various oracles, we propose in
this chapter to review the understanding we have achieved
about what constitutes a unipotent representation. There is
no claim that the definitions set forth here are completely
consistent with each other, or with what has gone before.
Chapters 7 through 11 ought to be regarded as support for
these definitions, however. All of them represent joint
work with Dan Barbasch.

We take G to be a reductive group in Harish–Chandra's
class.

Definition 12.1. Suppose that X is an irreducible $(\mathfrak{g},K)$-
module.

a) X is called special unipotent if its annihilator
in $U(\mathfrak{g})$ is a special unipotent primitive ideal.

b)   X   is called *distinguished unipotent* if its annihilator in   U($\mathfrak{g}$)   is a distinguished unipotent primitive
ideal.

c)   X   is called *unipotent* if its annihilator in   U($\mathfrak{g}$)
is a unipotent primitive ideal, and if certain other (as yet
not specified) conditions are satisfied (cf. Definition
11.25).

d)   X   is called *weakly unipotent* if its annihilator in
U($\mathfrak{g}$)   is a weakly unipotent primitive ideal.

The conditions "special unipotent" and "distinguished unipotent" should each imply unipotent; and unipotent should
imply weakly unipotent. No other relation among the conditions is true in general. We will define special, distinguished, and weakly unipotent primitive ideals; unipotent
primitive ideals are not yet defined (cf. Conjecture 9.19).
(With our definitions, special and distinguished unipotent
both imply weakly unipotent.)

To make the definitions, we need a little notation.
Fix a Cartan subalgebra $\mathfrak{h}$ of $\mathfrak{g}$, and define

(12.2) $\mathfrak{h}_{\mathbb{R}}^* = \mathbb{R}$-span of the roots of $\mathfrak{h}$ in $\mathfrak{g}$.

The invariant bilinear form $\langle \, , \, \rangle$ on $\mathfrak{g}$ is positive
definite on $\mathfrak{h}_{\mathbb{R}}^*$.

For λ in $\mathfrak{h}^*$, define

(12.3) $I(\lambda)$ = largest proper ideal in $U(\mathfrak{g})$

containing ker ξ_λ

(Definition 6.5); $I(\lambda)$ is a well-defined maximal ideal in

$U(\mathfrak{g})$.

Definition 12.4. A primitive ideal in $U(\mathfrak{g})$ is called

special unipotent if it is of the form $I(\lambda)$ (cf. (12.3)),

with λ arising as follows. Let ${}^d\mathfrak{g}$ be a semisimple Lie

algebra having a Cartan subalgebra ${}^d\mathfrak{h}$ isomorphic to $\mathfrak{h}^*$,

such that the coroots of ${}^d\mathfrak{h}$ in ${}^d\mathfrak{g}$ correspond to the

roots of $\mathfrak{h}$ in $\mathfrak{g}$. Then we require that there be a homo-

morphism ψ from $\mathfrak{sl}(2)$ to ${}^d\mathfrak{g}$, such that

$\psi(\mathrm{diag}(\tfrac{1}{2},-\tfrac{1}{2}))$ belongs to ${}^d\mathfrak{h}$, and corresponds

to the element λ of $\mathfrak{h}^*$.

By the Jacobson-Morozov theorem, special unipotent primitive

ideals are precisely parametrized by nilpotent orbits in the

dual semisimple Lie algebra ${}^d\mathfrak{g}$. If Ω is such an orbit,

then we write

(12.5) $\lambda_s(\Omega)$

for the corresponding weight.

Definition 12.6. A primitive ideal in $U(\mathfrak{g})$ is called *dis-

tinguished unipotent* if it is of the form $I(\lambda)$ (cf.

12.3)), with λ arising as follows. Fix a nilpotent orbit Λ in g^*. Call a weight μ in $\mathfrak{h}^*$ *acceptable for* Λ if there is an ideal I in $U(g)$ such that

1) I contains $\ker(\xi_\mu)$ (Definition 6.5); and

2) the variety of the associated graded ideal $gr(I)$ is the closure of Λ.

What we require of λ is

    i) λ is acceptable for Λ;

    ii) λ belongs to $h_{\mathbb{R}}^*$ (cf. 12.2); and

    iii) $\langle\lambda,\lambda\rangle$ is minimal subject to (i) and (ii).

A case-by-case calculation shows that λ exists, is unique, and satisfies

(12.7)(a)   variety of $gr(I(\lambda))$ = closure of Λ.

We may write

(12.7)(b) $\lambda = \lambda_d(\Lambda)$.

Consequently the distinguished unipotent primitive ideals are parametrized precisely by the nilpotent orbits in g^*.

Lack of Definition 12.8. A primitive ideal in $U(g)$ is called *unipotent* if it is of the form $I(\lambda)$ (cf. (12.3)), with λ arising as follows. Fix a nilpotent orbit Λ in g^*, and a (connected) covering $\tilde{\Lambda}$ of Λ. Then

$$\lambda = \lambda_u(\tilde{\Lambda})$$

is associated to $\Lambda^{\sim}$ in some still unspecified way (but cf.
Lemma 9.21 and Conjecture 9.18). If the covering is trivial, then we require

$$\lambda_u(\Lambda) = \lambda_d(\Lambda).$$

We will also require that every special unipotent primitive
ideal be unipotent. More precisely, recall from the appen-
dix to [Barbasch-Vogan, 1985] that Spaltenstein has defined
a map

$$\Omega \rightarrow {}^d\Omega$$

from nilpotent orbits in ${}^d\mathfrak{g}$ to (special) nilpotent orbits
in $\mathfrak{g}^*$. We require that

(12.9) $\lambda(\Omega) = \lambda_u(({}^d\Omega)^{\sim})$,

for some covering $({}^d\Omega)^{\sim}$ of ${}^d\Omega$. (The covering will depend
on Ω, and not just on ${}^d\Omega$.)

Definition 12.10. A primitive ideal I in $U(\mathfrak{g})$ is called
weakly unipotent if the following two conditions are satisfied. Choose λ in $\mathfrak{h}^*$ representing the infinitesimal
character of I (so that I contains the kernel of ξ_λ).
The first condition is

i) λ belongs to $\mathfrak{h}^*_{\mathbb{R}}$ (cf. (12.2)). Fix an irreducible representation X of $\mathfrak{g}$ with annihilator I. Consider the set of weights μ with the following property:

there is a finite dimensional representation F of $\mathfrak{g}$ such
that some non-zero vector in $X \otimes F$ is annihilated by the
kernel of ξ_μ (Definition 6.5). The second condition is

    ii) for every such μ, $\langle \mu, \mu \rangle \geq \langle \lambda, \lambda \rangle$.

Informally, the condition says that the infinitesimal character cannot be shortened by tensoring with a finite dimensional representation. This condition has an obvious
similarity to the definition of distinguished. It also
plays a significant part in the construction of other representations from unipotent ones (cf. Theorem 13.6).

Chapter 13

EXHAUSTION

In this chapter, we will offer a few hints about how

one can hope to prove that some list of irredcible unitary

representations is complete. The modern approach to this

problem begins with the *Langlands classification*. This is a

complete list of all irreducible (g,K)-modules, with expli-

citly described parameters; the representations themselves

are described in a slightly less explicit way. We stated

this theorem completely and carefully in the case of complex

G (Theorem 8.15). We will not do so for the general case,

but here is a part of the result.

THEOREM 13.1 ([Langlands, 1973]; see also [Borel-Wallach,

1980]). Suppose G is a reductive Lie group in Harish-

Chandra's class, and π is an irreducible admissible repe-

sentation of G on a Banach space V. Then we can find

i)   a parabolic subgroup $P = LN$ of G; and

ii)   an irreducible admissible representation γ of

L, with the following properties:

a)   the restriction of γ to the commutator sub-

group L' of L is a sum of discrete series representa-

tions of L' (Definition 5.10).

b) π is infinitesimally equivalent (Definition

2.14) to a subrepresentation of the induced representation

$$\text{Ind}_P^G(\gamma \otimes \mathbb{C}).$$

Much more is true, but this conveys the main idea. Theorem

8.15 suggests how much more detailed information is avail-

able.

Sketch of proof. Fix a non-zero continuous linear map p

from V onto a finite-dimensional representation (δ, E) of

K, intertwining the action of K.   Let $\mathcal{E}$ be the vector

bundle on G/K corresponding to E (Proposition 3.2). We

want to define a map from V^∞ (Definition 2.3) to the space

$C^\infty(G/K, \mathcal{E})$ of smooth sections of $\mathcal{E}$.   Given v in V, de-

fine a function f_w from G to E, by

$$f_v(g) = p(\pi(g^{-1})v).$$

(It is easy to show that f_v is smooth for any v in V,

but this fact is not needed.) Then

$$f_v(gk) = p(\pi(k^{-1})\pi(g^{-1})v)$$

$$= \delta(k^{-1})p(\pi(g^{-1})v)$$

$$= \delta(k^{-1})f_v(g).$$

By Corollary 3.4, f_v corresponds to a section F_v of $\mathcal{E}$. The map

(13.2)(a) $P\colon V^\infty \to C^\infty(G/K, \mathcal{E})$

defined in this way clearly intertwines the actions of G.

Suppose z is in the center of $U(\mathfrak{g})$. Write D_z for the corresponding right-invariant differential operator on G. Corollary 3.4 shows how to lift D_z to a differential operator (still called D_z) on sections of $\mathcal{E}$. By Lemma 6.6 and Theorem 2.12, z acts by some scalar $\xi_\lambda(z)$ on V^∞. By the intertwining property above, it follows that sections F in the image of P satisfy the differential equations

(13.2)(b) $(D_z - \xi_\lambda(z))F = 0 \quad (z \in \mathcal{Z}(\mathfrak{g})).$

At this point some serious analysis on G is needed. The idea is roughly that G/K can be compactified, by adding pieces at infinity that look like (among other things) various G/P's, with P parabolic. The prototypical example is $SU(1,1)$ acting on the closed unit disk. The interior is G/K, and the boundary circle is G/P. One use the differential equations (13.2)(b) to define boundary values of the sections F. Typical boundary values are sections of certain vector bundles on G/P; that is, they are vectors in induced representations. In this way one gets intertwining

operators from sections of $\mathcal{E}$ satisfying (13.2)(b), to in-

duced representations.

If these boundary value maps are non-zero, we are done.

If they are zero, the conclusion is that the sections of $\mathcal{E}$

under consideration tend to zero (in a well-controlled way)

at infinity. It follows that they are square-integrable,

and therefore (Definition 5.10) that π is a discrete

series representation. Then the theorem is still true, with

P equal to G.   □.

This outline is really more motivation than sketch;

there is no explicit compactification of G/K in

[Langlands, 1973], for example.

Recall from Definition 4.6 the notion of Hermitian dual

of a representation π . If π is an irreducible admissible

representation with specified parameters in the Langlands

classification, then one can compute the parameters of π^h

by a simple formal operation. In particular, one can deter-

mine easily from the Langlands parameters whether π admits

an invariant Hermitian form. Since we have not described

the Langlands parameters in detail, we cannot give the re-

sult (see [Knapp-Zuckerman, 1977]; but here is a special

case.

PROPOSITION 13.3 (see [Duflo, 1979]). *Suppose* G *is a complex connected reductive algebraic group,* H *is a Cartan subgroup of* G, *and* W *is the Weyl group of* H *in* G. *Let* π *be an irreducible admissible representation of* G, *corresponding to the character* χ *of* H *(Theorem 8.15). Then* π^h *corresponds to* χ^h, *the inverse of the complex conjugate of* χ. *In particular,* π *admits an invariant Hermitian form if and only if there is a* w *in* W *such that*

$$w\chi = \chi^h.$$

When w is 1 in this theorem, χ is a unitary character, and π is a unitarily induced representation. When w is -1, χ is real-valued.

A great deal is known about when induced representations like those in Theorem 13.1 can be reducible (see [Speh-Vogan, 1980], for example). This information, in conjunction with Theorem 13.1 and the formal analysis described above, leads to the following result.

Definition 13.4. **Suppose** $\mathfrak{g}$ **is a complex reductive Lie algebra with Cartan subalgebra** $\mathfrak{h}$. **We say that** ξ_λ **(Definition 6.5) is a** *real infinitesimal character* **if** λ **belongs to** $\mathfrak{h}^*_{\mathbb{R}}$ **(cf. (12.2)).**

THEOREM 13.5 (see [Knapp, 1986], Theorem 16.10). *Suppose* G
is a reductive Lie group in Harish-Chandra's class and π
is an irreducible admissible repesentation of G *on a*
Hilbert space $\mathcal{H}$. *Assume that* π *admits an invariant*
Hermitian form $\langle \, , \, \rangle_G$. *Then we can find*

i) *a parabolic subgroup* P = LN *of* G; *and*

ii) *an irreducible admissible representation* γ *of* L,
with the following properties:

a) *the restriction of* γ *to the commutator sub-*
group L' *of* L *has real infinitesimal character (Defini-*
tion 13.4);

b) γ *admits a non-degenerate invariant Hermitian*
form $\langle \, , \, \rangle_L$; *and*

c) π *is infinitesimally equivalent (Definition*
2.14) *to the induced representation*

$$\mathrm{Ind}_P^G(\gamma \otimes \mathbb{C}),$$

with the induced Hermitian form (defined in analogy with
(3.9)).

In particular, π *is unitary if and only if* γ *is.*

This result allows us to restrict attention to representations having real infinitesimal character. Roughly
speaking, such representations ought to come from the

derived functor construction of Theorem 6.8. One of the problems with that construction was that it did not always take unitary representations to unitary representations. We can now repair that problem to some extent.

THEOREM 13.6 (Theorem 7.1 and Proposition 8.17 in [Vogan, 1984]). *In the setting of Theorem 6.8, assume that* G *is in Harish-Chandra's class. Write* $\mathfrak{l}$ *as the direct sum of its center and its commutator subalgebra:*

$$\mathfrak{l} = c + \mathfrak{l}'.$$

Assume that

   i) *the restriction of* Z *to* $\mathfrak{l}'$ *is weakly unipotent (Definitions 12.1 and 12.10); and*

   ii) *the weight* λ_c *in* c^* *by which* c *acts on* Z *satisfies*

$$\text{Re } \langle \lambda_c, \alpha \rangle \geq 0$$

for any root α *of* c *in* $\mathfrak{u}$. *Then*

   a) $\mathcal{R}^j(Z)$ *is zero for not equal to* S;

   b) *any non-degenerate Hermitian form* $\langle \ , \ \rangle_L$ *on* Z *induces one* $\langle \ , \ \rangle_G$ *on* $\mathcal{R}^S(Z)$; *and*

   c) *if* $\langle \ , \ \rangle_L$ *is positive, then so is* $\langle \ , \ \rangle_G$.

The most pleasant feature of this result is that (if Z and L are fixed, L is a Levi factor for some θ-stable para-

bolic q', and Z is unitary on the center of L) hypothesis (ii) is always satisfied for at least one choice of
θ-stable parabolic q with Levi factor $\mathfrak{l}$. We therefore
have a rather complete way of passing from weakly unipotent
unitary representations of Levi factors, to unitary representations of G. The picture would be satisfactory indeed
if the following result were true.

FALSE THEOREM 13.7. *Suppose* G *is a reductive Lie group in
Harish-Chandra's class and* X *is an irreducible* $(\mathfrak{g},K)$-
*module having real infinitesimal character (Definition
13.4). Assume that* X *admits a positive definite invariant
Hermitian form* $\langle \ , \ \rangle_G$. *Then we can find*

i) a Levi subgroup L *of* G, *attached to a* θ-*stable
parabolic subalgebra* $q = \mathfrak{l} + \mathfrak{u}$; *and*

ii) an irreducible $(\mathfrak{l},(L \cap K)^{\sim})$-*module* Z, *with a
positive definite invariant Hermitian form* $\langle \ , \ \rangle_L$, *with the
following properties. Use the notation of Theorem 10.6.
Then*

a) the restriction of Z *to* $\mathfrak{l}'$ *is weakly unipotent;*

b) Z *admits a positive definite invariant
Hermitian form* $\langle \ , \ \rangle_L$;

c)   the weight λ_c in c^* by which   c   acts on

Z   satisfies

$$\langle \lambda , \alpha \rangle \geq 0$$

for any root α of   c   in $\mathfrak{u}$; and

d)   X   is isomorphic to $\mathscr{R}^S(Z)$, with the induced

Hermitian form.

The main reason this is false is that not all comple-

mentary series representations are weakly unipotent. (For

example, if   G   is $SL(2,\mathbb{R})$, the complementary series $C(\sigma)$

(Theorem 4.23) is weakly unipotent only for σ less than or

equal to   ½.)   Here is a weaker result, which follows from

[Vogan, 1984].

THEOREM 13.8   Suppose   G   is a semisimple group in Harish-

Chandra's class.   Then there is a computable constant c_G,

with the following property.   Suppose   X   is an irreducible

unitary $(\mathfrak{g},K)$-module, with real infinitesimal character ξ_λ

(Definition 13.4).   Assume that

$$\langle \lambda, \lambda \rangle \geq c_G.$$

Then we can find $\mathfrak{q}$,   L,   and   Z   as in Theorem 6.8, satisfy-

ing

a) $\mathfrak{q}$ is not equal to $\mathfrak{g}$;

b) Z is unitary and irreducible;

c) the positivity hypothesis of Theorem 6.8(d) holds; and

d) X is isomorphic to $\mathscr{R}^S(Z)$.

Sketch of proof. Using Theorem 13.1, write X as a subrepresentation of some induced representation. Use the notation established for Theorem 4.11; in our case, ξ is a discrete series representation of M. Then

(13.9)(a) $$X \subset I_p(\xi \otimes v).$$

Since X has real infinitesimal character, v is real-valued.

The analysis described before Proposition 13.3 shows that the Hermitian form on X arises as follows: there is an element w of $W(G,MA)$, of order 2, such that

(13.9)(b) $$w \cdot \xi = \xi, \quad w \cdot v = -v.$$

The Hermitian form on X is given by

(13.9)(c) $$\langle v_1, v_2 \rangle^X = \langle v_1, A(w \colon v)v_2 \rangle^h.$$

Here v_1 and v_2 are elements of the space $\mathscr{H}$ of the induced representation that belong to the subspace X, and $\langle \ , \ \rangle^h$ is the inner product on $\mathscr{H}$ given by integration over $K/(K \cap P)$.

The proof proceeds by analyzing the form

(13.10) $$\langle v_1, v_2 \rangle^t = \langle v_1, A(w \colon tv)v_2 \rangle^h$$

on $\mathcal{H}$, as a function of the real variable t. Theorem 4.11 guarantees that it is meromorphic in t, with poles and zeros only when $I_p(\xi \otimes t\upsilon)$ is reducible. When t is zero, the induced representation is unitary, and its reducibility is precisely known; the signature of $\langle \, , \, \rangle^0$ can be calculated exactly. The way that the signature varies with t is controlled by the reducibility of the induced representation.

On the other hand, [Speh–Vogan, 1980] allows one to describe the reducibility of $I_p(\xi \otimes t\upsilon)$ in terms of an analogous problem on some Levi factor L of a θ–stable parabolic, as long as $t\upsilon$ is not too large compared to the infinitesimal character λ_M of ξ. (This is the most difficult step in the argument; I will not discuss the ideas involved. They are one of the main topics of [Vogan, 1981].) The description is implemented by the cohomological induction functor $\mathcal{R}^S$. The conclusion of the theorem now drops out, as long as υ is not too large compared to λ_M.

On the other other hand, the proof of Theorem 13.1 produces υ from X in terms of the behavior of matrix coefficients of X at infinity. If X is unitary, its matrix coefficients must be bounded; so the corresponding υ cannot be too big. The infinitesimal character of X is that of the induced representation, which is (λ_M, υ). Since

X is assumed to have large infinitesimal character, this forces λ_M to be large. Now v is not too large compared to λ_M, as desired.   □

This argument is replete with information even when X has small infinitesimal character; many of the recent results on classifying unitary representations are based in part on it.

REFERENCES

J. Arthur, "Eisenstein series and the trace formula," in *Automorphic Forms, Representations, and L-functions,* Proceedings of Symposia in Pure Mathematics **33**, part 2, 27–61. American Mathematical Society, Providence, Rhode Island, 1979.

J. Arthur, "On some problems suggested by the trace formula," in *Proceedings of the Special Year in Harmonic Analysis, University of Maryland,* R. Herb, R. Lipsman, and J. Rosenberg, eds. Lecture Notes in Mathematics **1024**, Springer–Verlag, Berlin–Heidelberg–New York, 1983.

M. Atiyah and W. Schmid, "A geometric construction of the discrete series for semisimple Lie groups," Invent. Math. **42** (1977), 1–62.

L. Auslander and B. Kostant, "Polarization and unitary representations of solvable Lie groups," Invent. Math. **14** (1971), 255–354.

M. W. Baldoni-Silva and A. Knapp, "Indefinite intertwining operators," Proc. Nat. Acad. Sci. U.S.A. **81** (1984), 1272–1275.

D. Barbasch and D. Vogan, "The local structure of characters," J. Funct. Anal. **37** (1980), 27–55.

D. Barbasch and D. Vogan, "Unipotent representations of complex semisimple Lie groups," Ann. of Math. **121** (1985), 41–110.

V. Bargmann, "Irreducible unitary representations of the Lorentz group," Ann. of Math. **48** (1947), 568–640.

A. Borel and N. Wallach, *Continuous Cohomology, Discrete Subgroups, and Representations of Reductive Groups.* Princeton University Press, Princeton, New Jersey, 1980.

R. Bott, "Homogeneous vector bundles," Ann. of Math. **66** (1957), 203–248.

F. Bruhat, "Sur les représentations induites des groupes de Lie," Bull. Soc. Math. France **84** (1956), 97–205.

N. Conze, "Quotients primitifs des algèbres enveloppantes et algèbres d'opérateurs différentiels," C. R. Acad. Sci. Paris Sér A-B 277 (1973), A1033-A1036.

P. Deligne and G. Lusztig, "Representations of reductive groups over finite fields," Ann. of Math. 103 (1976), 103-161.

J. Dixmier, Algèbres Enveloppantes. Gauthier-Villars, Paris-Brussels-Montreal 1974.

J. Dixmier, Von Neumann Algebras, translated by F. Jellett. North-Holland Publishing Co., New York, 1981.

M. Duflo, "Représentations irréductibles des groupes semi-simples complexes," in Analyse Harmonique sur les Groupes de Lie (Sem. Nancy-Strasbourg 1973-75). Lecture Notes in Mathematics 497, Springer-Verlag, Berlin-Heidelberg-New York, 1975.

M. Duflo, "Représentations unitaires irréductibles des groupes semi-simples complexes de rang deux," Bull. Soc. Math. France 107 (1979), 55-96.

M. Duflo, "Construction de représentations d'un groupe de Lie," Cours d'été du C.I.M.E., Cortona 1980.

M. Duflo, "Théorie de Mackey pour les groupes de Lie algèbriques," Acta Math. 149 (1982), 153-213.

M. Duflo, "Construction de gros ensembles de représentations unitaires irréductible d'un groupe de Lie quelconque," in Operator Algebras and Group Representations, volume I. Monographs Stud. Math. 17. Pitman, Boston, Massachusetts-London, 1984.

T. J. Enright and N. R. Wallach, "Notes on homological algebra and representations of Lie algebras," Duke Math. J. 47 (1980), 1-15.

I. M. Gelfand and M. A. Naimark, "Unitary representations of the Lorentz group," Izv. Akad. Nauk S.S.S.R. 11 (1947), 411-504.

V. Guillemin and S. Sternberg, Geometric Asymptotics, Math Surveys 14. American Mathematical Society, Providence, Rhode Island (1978).

Harish-Chandra, "Representations of semi-simple Lie groups
I," Trans. Amer. Math. Soc. **75** (1953), 185-243.

Harish-Chandra, "Spherical functions on a semisimple Lie
group II," Amer. J. Math. **80** (1958), 553-613.

Harish-Chandra, "Invariant eigendistribtions on a semisimple Lie group," Trans. Amer. Math. Soc. **119** (1965),
457-508.

Harish-Chandra, "Discrete series for semi-simple Lie groups
II," Acta Math. **116** (1966), 1-111.

S. Helgason, *Differential Geometry, Lie Groups, and Symmetric Spaces*. Academic Press, New York, San Francisco,
London, 1978.

S. Helgason, *Groups and Geometric Analysis*. Academic
Press, Orlando, Florida, 1984.

R. Howe, "On the role of the Heisenberg group in harmonic
analysis," Bull. Amer. Math. Soc. (N.S.) **3** (1980),
821-843.

J. E. Humphreys, *Introduction to Lie Algebras and Representation theory*. Springer-Verlag, Berlin-Heidelberg-New
York, 1972.

J. C. Jantzen, *Moduln mit einem Höchsten Gewicht*, Lecture
Notes in Mathematics **750**. Springer-Verlag, Berlin-
Heidelberg-New York, 1979.

A. Joseph, "The minimal orbit in a simple Lie algebra and
its associated maximal ideal," Ann. Scient. Ec. Norm.
Sup. (4) **9** (1976), 1-30.

A. Joseph, "Goldie rank in the enveloping algebra of a semisimple Lie algebra I, II," J. Alg. **65** (1980), 269-316.

A. Joseph, "Kostant's problem and Goldie rank," 249-266, in
Non-commutative Harmonic Analysis and Lie Groups, J.
Carmona and M. Vergne, eds., Lecture Notes in Mathematics **880**. Springer-Verlag, Berlin-Heidelberg-New York,
1981.

A. Joseph, "On the classification of primitive ideals in
the enveloping algebra of a semisimple Lie algebra," in

Lie group representations I, R. Herb, R. Lipsman and J. Rosenberg, eds. Lecture Notes in Mathematics **1024**. Springer-Verlag, Berlin-Heidelberg-New York, 1983.

D. Kazhdan and G. Lusztig, "Representations of Coxeter groups and Hecke algebras," Inventiones Math. **53** (1979), 165-184.

A. Kirillov, "Unitary representations of nilpotent Lie groups," Uspehi Mat. Nauk. **17** (1962), 57-110.

A. Knapp, *Representation Theory of Real Semisimple Groups: An Overview Based on Examples.* Princeton University Press, Princeton, New Jersey, 1986.

A. Knapp and B. Speh, "The role of basic cases in classification: theorems about unitay representations applicable to SU(N,2)," 119-160 in *Non-commutative Harmonic Analysis and Lie Groups*, J. Carmona and M. Vergne, eds., Lecture Notes in Mathematics **1020**. Springer-Verlag, Berlin-Heidelberg-New York-Tokyo, 1983.

A. Knapp and E. Stein, "Singular integrals and the principal series. IV," Proc. Nat. Acad. Sci. U.S.A. **72** (1975), 2459-2461

A. Knapp and E. Stein, "Intertwining operators for semisimple groups II," Inventiones Math. **60** (1980), 9-84.

A. Knapp and D. Vogan, "Duality theorems in relative Lie algebra cohomology," preprint, 1986.

A. Knapp and G. Zuckerman, "Classification theorems for representations of semisimple Lie groups," 138-159 in *Non-commutative Harmonic Analysis*, J. Carmona and M. Vergne, eds., Lecture Notes in Mathematics **587**. Springer-Verlag, Berlin-Heidelberg-New York, 1977.

B. Kostant, "The principal three-dimensional subgroup and the Betti numbers of a complex simple Lie group," Amer. J. Math. **81** (1959), 973-1032.

B. Kostant, "On the existence and irreducibility of certain series of representations," Bull. Amer. Math. Soc. **75** (1969), 627-642.

B. Kostant, "Coadjoint orbits and a new symbol calculus for line bundles," in *Conference on Differential Geometric*

Methods in Theoretical Physics (G. Denardo and H.D. Doebner, eds.). World Scientific, Singapore, 1983.

B. Kostant and S. Rallis, "Orbits and representations associated with symmetric spaces," Amer. J. Math. 93 (1971), 753–809.

S. Lang, *SL(2,$\mathbb{R}$)*. Addison-Wesley, Reading, Massachusetts, 1975.

R. P. Langlands, "On the classification of representations of real algebraic groups," preprint, Institute for Advanced Study, 1973.

J. Lepowsky, "Algebraic results on representations of semi-simple Lie groups," Trans. Amer. Math. Soc. 176 (1973), 1–44.

G. Lusztig, *Characters of Reductive Groups over a Finite Field*. Annals of Mathematics Studies 107. Princeton University Press, Princeton, New Jersey, 1984.

G. Mackey, *Theory of Unitary Group Representations*. University of Chicago Press, Chicago, 1976.

F.I. Mautner, "Unitary representations of locally compact groups II," Ann. of Math 52 (1951), 528–556.

E. Nelson, "Analytic vectors," Ann. of Math. 70 (1959), 572–615.

H. Ozeki and M. Wakimoto, "On polarizations of certain homogeneous spaces," Hiroshima Math. J. 2 (1972), 445–482.

F. Peter and H. Weyl, "Die Vollständigkeit der primitiven Darstellungen einer geschlossenen kontinuierlichen Gruppe," Math. Ann. 97 (1927), 737–755.

Pontriagin, *Topological Groups*, translated by E. Lehmer. Princeton University Press, Princeton, New Jersey, 1939.

J. Rawnsley, W. Schmid, and J. Wolf, "Singular unitary representations and indefinite harmonic theory," J. Funct. Anal. 51 (1983), 1–114.

W. Shakespeare, A *Midsummer-Night's Dream*, act V, scene i, lines 12-17.

W. Schmid, "Homogeneous complex manifolds and representations of semisimple Lie groups," Ph.D. dissertation, University of California at Berkeley, Berkeley, 1967.

W. Schmid, "Some properties of square-integrable representations of semisimple Lie groups," Ann. of Math. **102** (1975), 535-564.

W. Schmid, "L_2 cohomology and the discrete series," Ann. of Math. **103** (1976), 375-394.

D. Shale, "Linear symmetries of free boson fields," Trans. Amer. Math. Soc. **103** (1962), 149-167.

B. Speh, "Unitary representations of $SL(n, \mathbb{R})$ and the cohomology of congruence subgroups," 483-505 in *Non-commutative Harmonic Analysis and Lie groups*, J. Carmona and M. Vergne, eds., Lecture Notes in Mathematics **880**. Springer-Verlag, Berlin-Heidelberg-New York, 1981.

B. Speh and D. Vogan, "Reducibility of generalized principal series representatons," Acta Math. **145** (1980), 227-299.

T. A. Springer, "Reductive groups," in *Automorphic Forms, Representations, and L-functions*, Proceedings of Symposia in Pure Mathematics **33**, part 1, 3-28. American Mathematical Society, Providence, Rhode Island, 1979.

E. M. Stein, "Analysis in matrix spaces and some new representations of $SL(n, \mathbb{R})$," Ann. of Math. **86** (1967), 461-490.

M. J. Taylor, *Noncommutative Harmonic Analysis*. American Mathematical Society, Providence, Rhode Island, 1986.

P. Torasso, "Quantification geometrique, operateurs d'entrelacement et representations unitaires de $SL_3(\mathbb{R})^\sim$," Act Math. **150** (1983), 153-242.

D. Vogan, "Gelfand-Kirillov dimension for Harish-Chandra modules," Inventiones math. **48** (1978), 75-98.

REFERENCES

D. Vogan, *Representations of Real Reductive Lie Groups*. Birkhauser, Boston-Basel-Stuttgart, 1981.

D. Vogan, "Unitarizability of certain series of representations," Ann. of Math. 120 (1984), 141-187.

D. Vogan, "The orbit method and primitive ideals for semisimple Lie algebras," in *Lie Algebras and Related Topics*, CMS Conference Proceedings, volume 5, D. Britten, F. Lemire, and R. Moody, eds. American Mathematical Society for CMS, Providence, Rhode Island, 1986.

D. Vogan, "The unitary dual of GL(n) over an archimedean field," Invent. Math. 83 (1986), 449-505.

J. von Neumann, "Die eindeutigkeit der Schröderschen Operatoren," Math. Ann. 104 (1931), 570-578.

N. Wallach, *Harmonic Analysis on Homogeneous Spaces*. Marcel Dekker, New York, 1973.

N. Wallach, "On the unitarizability of derived functor modules," Invent. Math. 78 (1984), 131-141.

G. Warner, *Harmonic Analysis on Semisimple Lie Groups*. Springer-Verlag, Berlin, Heidelberg, New York, 1972.

A. Weil, *L'intégration dans les Groupes Topologiques et ses Applications*. Actual. Sci. Ind. 869. Hermann et Cie., Paris, 1940.

H. Weyl, "Theorie der Darstellung kontinuierlicher Gruppen. I, II, III," Math. Z. 23 (1925), 271-309; 24 (1925), 328-376, 377-395.

E. Wigner, "On unitary representations of the inhomogeneous Lorentz group," Ann. of Math. 40 (1939), 149-204.

Library of Congress Cataloging-in-Publication Data

Vogan, David A., 1954-
   Unitary representations of reductive Lie groups.

   (Annals of mathematics studies ; no. 118)
   Bibliography: p.
   1. Lie groups.   2. Representations of groups.
I. Title.   II. Series.
QA387.V64 1987   512'.55   87-3102
ISBN 0-691-08481-5
ISBN 0-691-08482-3 (pbk.)